ENDOCRINOLOGY RESEARCH AND CLINICAL DEVELOPMENTS

REGUCALCIN

METABOLIC REGULATION AND DISEASE

ENDOCRINOLOGY RESEARCH AND CLINICAL DEVELOPMENTS

Additional books and e-books in this series can be found on Nova's website under the Series tab.

ENDOCRINOLOGY RESEARCH AND CLINICAL DEVELOPMENTS

REGUCALCIN

METABOLIC REGULATION AND DISEASE

MASAYOSHI YAMAGUCHI
EDITOR

Copyright © 2019 by Nova Science Publishers, Inc.

All rights reserved. No part of this book may be reproduced, stored in a retrieval system or transmitted in any form or by any means: electronic, electrostatic, magnetic, tape, mechanical photocopying, recording or otherwise without the written permission of the Publisher.

We have partnered with Copyright Clearance Center to make it easy for you to obtain permissions to reuse content from this publication. Simply navigate to this publication's page on Nova's website and locate the "Get Permission" button below the title description. This button is linked directly to the title's permission page on copyright.com. Alternatively, you can visit copyright.com and search by title, ISBN, or ISSN.

For further questions about using the service on copyright.com, please contact:
Copyright Clearance Center
Phone: +1-(978) 750-8400   Fax: +1-(978) 750-4470   E-mail: info@copyright.com.

NOTICE TO THE READER

The Publisher has taken reasonable care in the preparation of this book, but makes no expressed or implied warranty of any kind and assumes no responsibility for any errors or omissions. No liability is assumed for incidental or consequential damages in connection with or arising out of information contained in this book. The Publisher shall not be liable for any special, consequential, or exemplary damages resulting, in whole or in part, from the readers' use of, or reliance upon, this material. Any parts of this book based on government reports are so indicated and copyright is claimed for those parts to the extent applicable to compilations of such works.

Independent verification should be sought for any data, advice or recommendations contained in this book. In addition, no responsibility is assumed by the Publisher for any injury and/or damage to persons or property arising from any methods, products, instructions, ideas or otherwise contained in this publication.

This publication is designed to provide accurate and authoritative information with regard to the subject matter covered herein. It is sold with the clear understanding that the Publisher is not engaged in rendering legal or any other professional services. If legal or any other expert assistance is required, the services of a competent person should be sought. FROM A DECLARATION OF PARTICIPANTS JOINTLY ADOPTED BY A COMMITTEE OF THE AMERICAN BAR ASSOCIATION AND A COMMITTEE OF PUBLISHERS.

Additional color graphics may be available in the e-book version of this book.

Library of Congress Cataloging-in-Publication Data

ISBN: 978-1-53616-172-4
Library of Congress Control Number:2019948413

Published by Nova Science Publishers, Inc. † New York

CONTENTS

Preface		vii
Acknowledgments		xiii
Chapter 1	Structure of Regucalcin Gene and Protein *Noriaki Shimokawa*	1
Chapter 2	The Horizons of Regucalcin in Male Reproduction *Sara Correia, Harikrishna Pillai,* *Ana Manuela S. Silva, Vanessa M. Santos* *and Silvia Socorro*	15
Chapter 3	Regucalcin: A Tumor-Associated Antigen and a Tumor-Suppressing Molecule *Su-Fang Zhou*	51
Chapter 4	The Role of Regucalcin in Bone Remodeling and Osteoporosis *Masayoshi Yamaguchi*	79
Chapter 5	Involvement of Regucalcin in Diabetes and Hyperlipedemia *Masayoshi Yamaguchi*	103

Chapter 6	Extracellular Vesicles: A Conversation in the Body *Naoomi Tominaga*	**121**
About the Editor		**145**
Index		**147**
Related Nova Publications		**159**

PREFACE

This book, which is entitled "Regucalcin: Metabolic Regulation and Disease", introduces the recent topics regarding the role of regucalcin in metabolic regulation and its related diseases. Looking back, there was a potent interest for biomedical research in the fields of cell signaling linked to hormone action. In 1970s, cyclic AMP and calcium (Ca^{2+}), which play a pivotal role as second messenger, were a major molecule of interest that contributes to signal transduction in the regulation of cellular function by peptide hormones. Afterwards, calmodulin was discovered as a modulator protein of intracellular Ca^{2+} signaling in hormonal action. In this context, protein kinase C, which is activated by Ca^{2+} and phospholipid, was also found to be a key protein in signal transduction. The discovery of these proteins leaded to establishing of the molecular mechanism of intracellular signaling system in various types of cells and tissues. After that, manifold proteins and its related signaling molecules were demonstrated to participate in novel signaling pathways related to various cytokines in cells. Systemic mechanism by which suppresses signaling process linked to the regulation of transcription activity was remained to elucidate.

Regucalcin was discovered in 1978 as a novel Ca^{2+}-binding protein not including the EF-hand motif, a specific site of Ca^{2+}-binding, which was found in various Ca^{2+}-binding proteins [Chemical & Pharmaceutical Bulletin (*Tokyo*), 26: 1915-1918, 1978 and 36: 321-325, 1988]. Notably, regucalcin was found to inhibit stimulatory effects of Ca^{2+}, calmodulin and protein kinase C on various enzyme activities in various types of cells. Furthermore, regucalcin was demonstrated to play a potential role as a suppressor in intracellular signaling systems linked to regulation of transcription activity. After manifold subsequent studies, regucalcin has been demonstrated to play a multifunctional role in the regulation of the function of various types of cells and tissues; it plays a pivotal role in the regulation of intracellular Ca^{2+} homeostasis, activity of various enzymes, cell signal transduction, nuclear function and gene expression, cell proliferation, apoptosis, and metabolic events. Regucalcin is proposed to play an essential role in maintaining cell homeostasis and protecting from disorders in various types of cells and tissues involved in aging.

This book focuses recent highlighted information regarding the role of regucalcin in metabolic regulation and its related diseases with aging. The research of regucalcin may contribute to development of biomedical sciences in feature. This book is constituted of chapters 1-6, and those key messages are summarized in the following.

Chapter 1 summarizes recent understanding about the structure of regucalcin gene and protein. In human, the regucalcin gene is present in the X chromosome and its length is approximately 15 kb. It consists of eight exons. At least three isoforms of regucalcin have been identified to date. All isoforms are biosynthesized by alternative splicing of exons. Amino acid sequence of regucalcin was predicted from its nucleotide sequence. The cDNA and amino acid sequences of regucalcin showed no significant homology with any of the other gene/protein sequences. The homology is high when comparing the amino acid sequences of humans and other mammalian regucalcin but not high between humans and animals other than mammals. Regarding the structural feature of regucalcin protein, the regucalcin has a 6-bladed β-propeller fold that contains the binding site of a single metal ion with either a Ca^{2+} or a Zn^{2+} in the center. Furthermore, when this molecule acts as a gluconolactonase, the cavity, formed by the 11 amino acids conserved in many mammals and their configuration, is bound to the substrate.

Chapter 2 emphasizes the horizons of regucalcin in male reproduction. Regucalcin is a highly conserved calcium (Ca^{2+})-binding protein, present throughout the evolution line, from prokaryotes to eukaryotes. It is highly expressed in the liver and kidney of mammalian vertebrates, but its broad expression in a panoply of other tissues has been reported. Over the years, the role of regucalcin in the regulation of several biological processes apart from Ca^{2+} homeostasis also was being disclosed. Regucalcin was shown to be a multifunctional protein involved in the regulation of intracellular signaling pathways, oxidative stress, cell proliferation, apoptosis, and also energetic metabolism. All these are crucial processes for the successful production of male gametes, which propelled researchers investigating the role of regucalcin in male reproduction. This chapter systematically describes the existent literature on the actions of regucalcin supporting spermatogenesis and sperm function, and also discusses the recent experimental advances that sustain the usefulness of this protein in reproductive technology.

Chapter 3 provides highlighted findings of regucalcin as a tumor-associated antigen and a tumor-suppressing molecule. Regucalcin is a multifunctional molecule and various studies have suggested that it is closely related to the development of tumors. It has been found that anti-regucalcin antibodies are present in the serum of some tumor patients such as liver cancer sufferers. The anti-regucalcin antibody positive rate is higher in α-fetoprotein-negative liver cancer serum than that in α-fetoprotein-positive ones. The frequency of serum antibody against regucalcin with well histopathologically differentiated samples is significantly higher than that in poorly differentiated ones. This indicates that regucalcin is a tumor-associated antigen and it can be used as a serum marker for tumor diagnosis. Regucalcin is expressed in normal tissues in the body, especially in liver and kidney. It is, however, generally showing low expression in tumor tissues, which is closely related to the shorter survival rate of patients. The low expression level of regucalcin in liver cancer cells and liver cancer tissues is inversely proportional to the frequency of methylation of its gene promoter. *In vitro* experiments have shown that silencing regucalcin in tumor cells will promote tumor cell invasion and migration. Regucalcin inhibits the proliferation of tumor cells by inhibiting the expression of multiple protein kinases, increasing the expression of tumor suppressor genes, reducing oncogene expression and regulating apoptosis-related proteins with independent of cell death. Those evidences suggest that regucalcin is a tumor-suppressing molecule and its level of expression can be used as a prognostic indicator for patients.

Chapter 4 discusses the role of regucalcin in the regulation of bone remodeling and osteoporosis. Bone homeostasis is regulated by the functions of osteoclast and osteoblasts through various hormones and cytokines in systemic and microenvironment of bone marrow cells. Overexpressed regucalcin induces bone loss related to osteoporosis in regucalcin transgenic rats *in vivo* and deficiency of regucalcin causes osteomalacia with impairment of endogenous vitamin C synthesis *in vivo*. Regucalcin mRNA and its protein are expressed in the femoral tissues, bone marrow cells, and osteoblastic cells. Extracellular regucalcin stimulates osteoclastogenesis in bone marrow culture *in vitro* and *in vitro* and exhibits suppressive effects on the differentiation and mineralization in osteoblastic cells. Moreover, extracellular regucalcin was found to suppress osteoblastogenesis and stimulate adipogenesis in bone marrow culture system *ex vivo*. Regucalcin reveals enhancing effects on activation of nuclear factor kapper B, which is mediated through tumor necrosis factor-α or the receptor activator of NF-κB ligand in preosteoblastic cells and preosteoclastic cells. Regucalcin plays a pivotal role in the regulation of bone homeostasis as a suppressor in osteoblastogenesis and an enhancer in osteoclastogenesis, demonstrating a role as a cytokine.

Chapter 5 raises the involvement of regucalcin in diabetes and hyperlipidemia. Regucalcin, which plays a multifunctional role in the regulation in various types and tissues, may play a pathophysiological role in metabolic disorder. The expression of regucalcin is stimulated through action of insulin in liver cells *in vitro* and *in vivo,* and it is diminished in the liver of rats with type I diabetes after streptozotocin administration *in vivo*. Overexpressed regucalcin stimulates glucose utilization and lipid production in liver cells with glucose supplementation *in vitro*. Regucalcin exhibits insulin resistance in liver cells. Deficiency of regucalcin induces an impairment of glucose tolerance and lipid accumulation in the liver of mice *in vivo*. Overexpressed regucalcin stimulates glucose utilization and lipid production in modeled rat hepatoma H4-II-E cells. Overexpressed regucalcin enhances glucose transporter 2 mRNA expression to stimulate glucose utilization, and it depresses the gene expression of insulin receptor or PI3 kinase involved in insulin signaling, which is enhanced by insulin and/or glucose supplementation, leading to insulin resistance in liver cells. Furthermore, overexpressed regucalcin decreases in triglyceride, total cholesterol and glycogen contents in the liver of rats *in vivo*, inducing a hyperlipidemia. Leptin and adiponectin mRNA expressions in the liver tissues are suppressed in regucalcin transgenic rats. The decrease in hepatic regucalcin is associated with the development and progression of nonalcoholic fatty liver disease and fibrosis in human patients. Regucalcin may be a key molecule implicated in diabetes and lipid metabolic disorder.

Chapter 6 proposes the recent topics regarding the role of regucalcin in extracellular vesicles and its related diseases. Through about four decades, many researchers had been uncovering the function of the extracellular vesicle. Extracellular vesicles, which is a nano mater scale vesicle, are made as multivesicular bodies in cells through early endosome. These are secreted from multivesicular bodies fusion with the plasma membrane. It is contained many functional molecules such as DNAs, mRNAs, microRNAs, and proteins, including regucalcin. Moreover, the extracellular vesicles are also contained in breast milk. These molecules are transferred from the donor cells to the other recipient cells to change the environment and/or to send some signals from cell to cell, which seems like a conversation of the cells. The extracellular vesicles are related to the malignancy of cancer, the development of neuron, and the immune system. Regucalcin also found in the extracellular vesicles from urine, which are downregulated in the state of the diabetic nephropathy. The extracellular vesicles were named as "Exosome" at first as the definition of "the released vesicles which may serve a physiologic function". However, the extracellular vesicles are named very different terminology through about four decades. The terminology is confusing because of the different correction methods in recent year. In this chapter, it will be an overview the history, terminology, collection methods, contents and function of extracellular vesicles, and involvement of regucalcin.

As introduced above, this book focuses recent topics on the pivotal role of regucalcin in metabolic regulation and diseases. The editor believes that this book will be of interest to graduate students, researcher, scientists and physicians that are focused on the fields of molecular and cellular biology, biomedical sciences, and clinical challenges. The editor hopes that the research of regucalcin will greatly contribute to the development of biomedical sciences in complication with various diseases and will be further developed in the feature to still cover a wide area of molecular, cellular, physiological, biomedical and clinical aspects.

Editor
Masayoshi Yamaguchi, PhD, IOM, FAOE, DDG, DG
Visiting Professor
Department of Pathology and Laboratory Medicine,
David Geffen School of Medicine,
University of California, Los Angeles (UCLA),
Los Angeles, USA

ACKNOWLEDGMENTS

This book was reviewed by:
Tomiyasu Murata, Ph.D.
Associate Professor
Faculty of Pharmacy
Meijo University, Nagoya, Japan

In: Regucalcin
Editor: Masayoshi Yamaguchi

ISBN: 978-1-53616-172-4
© 2019 Nova Science Publishers, Inc.

Chapter 1

STRUCTURE OF REGUCALCIN GENE AND PROTEIN

Noriaki Shimokawa[*]
Department of Nutrition,
Takasaki University of Health and Welfare, Gunma, Japan

ABSTRACT

This chapter aims to summarize recent understanding about the structure of regucalcin (RGN) gene and protein. In human, RGN gene is present in the X chromosome and its length is approximately 15 kb. It consists of eight exons. At least three isoforms of RGN have been identified to date. All isoforms are biosynthesized by alternative splicing of exons. Amino acid sequence of RGN was predicted from its nucleotide sequence. The cDNA and amino acid sequences of RGN showed no significant homology with any of the other gene/protein sequences. The homology is high when comparing the amino acid sequences of humans and other mammalian RGN but not high between humans and animals other than mammals. Regarding the structural feature of RGN protein, the RGN has a 6-bladed β-propeller fold that contains the binding site of a single metal ion with either a Ca^{2+} or a Zn^{2+} in the center. Furthermore,

---
[*] Corresponding Author's E-mail: shimokawa-n@takasaki-u.ac.jp.

when this molecule acts as a gluconolactonase, the cavity, formed by the 11 amino acids conserved in many mammals and their configuration, is bound to the substrate.

Keywords: regucalcin, RGN, Ca^{2+}-binding protein, cDNA and genomic cloning, structural features of protein

INTRODUCTION

Regucalcin (RGN) was found and named by Yamaguchi in 1978 [1, 2]. Around the time of its discovery, RGN was determined to be a calcium-binding protein [3]. It has since been reported that the Ca^{2+} binding constant of RGN is 4.19×10^5 M^{-1} according to equilibrium dialysis [4]. Furthermore, the intraperitoneal administration of calcium chloride induced a remarkable increase in RGN mRNA in liver [5], and its expression was shown to be mediated through Ca^{2+}/calmodulin [6]. Recently, however, RGN has attracted the interest of many researchers as a multi-functional protein [7–9]. RGN is sometimes referred to as senescence marker protein-30 (SMP30), which was found by another group for a different research purpose from RGN after the discovery of RGN [10, 11]. This chapter describes the gene analysis findings and its product (protein) characteristics that are the basis of the functional diversity of RGN.

CLONING OF RGN CDNA

Complementary DNA (cDNA) of RGN was cloned by Shimokawa and Yamaguchi in 1993 [12]. After the extraction of total RNA from the liver of male Wistar rats (3-week-old), poly(A)$^+$ RNA was purified using an oligo(dT)-cellulose column. Double-stranded cDNA was then synthesized using mRNA as a template. cDNA was ligated with the phage expression vector a λZAP II and subsequently packaged into native phages using *in vitro* packaging extracts. Using rabbit anti-RGN antiserum [13], clones that contained a complete open reading frame of RGN were isolated by immunoscreening from 1×10^6 plaques.

Nucleotide sequencing was performed by the dideoxynucleotide termination method. The translation start site (ATG codon) of amino acids of RGN was identified based on the Kozak sequence (GCCA/GCC<u>ATG</u>G) [14]. The number of amino acids in RGN as predicted from the nucleotides was 299, and the molecular mass was calculated to be 33.4 kD. On using a computer-assisted alignment search tool to perform a homology search (FASTA and BLAST, NCBI), the cDNA and amino acid sequences of RGN showed no significant homology with any of the other gene/protein sequences registered in EMBL or GenBank at the time.

CHROMOSOMAL LOCALIZATION AND EXON ALIGNMENT OF THE RGN GENE

The chromosomal localization of the RGN gene was experimentally determined by the direct R-banding fluorescence *in situ* hybridization (FISH) method for the first time in rats [15]. It is present in the X chromosome of some mammals, including rats (Xq11.1-12) and humans (Xp12) (Figure 1A). The length of the human RGN gene is approximately 15 kb (https://www.ncbi.nlm.nih.gov/nuccore/NC_000023.11?report=genbank&from=47078355&to=47093314). It consists of eight exons, as shown in Figure 1B [15, 16]. Exons 1 and 2 are non-coding regions (5' UTR), and translation (CDS) starts from a position near the end of exon 3. All exons 4 to 7 are translated. A stop codon (tga) is located near the start of exon 8, and the translation is terminated. It then continues to the non-translated region (3' UTR). The start of the intron of the human RGN gene is GT, and the end is AG, following the general GT-AG rule (Chambon's rule) [17].

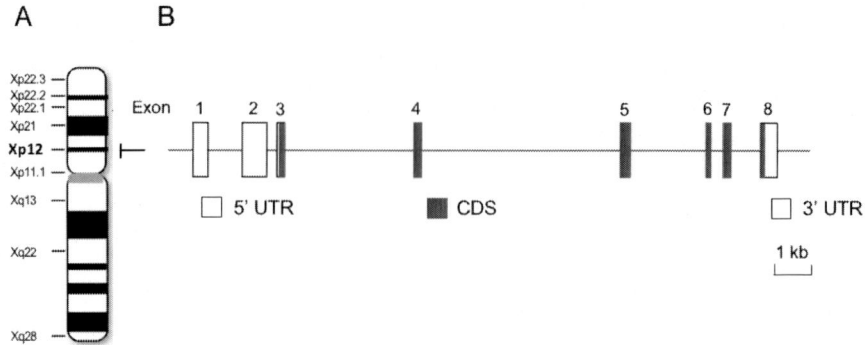

Figure 1. Chromosomal localization and exon alignment of the RGN gene. A. The RGN gene is localized to human chromosome Xp12. B. The genomic organization of the RGN gene. The positions of exons are shown as boxes (1 – 8). The coding sequence (CDS) and 5'- and 3'-untranslated regions are shown as solid-red squares (■) and open squares (□), respectively.

ISOFORMS OF RGN

At least three isoforms of RGN have been identified to date (https://www.ncbi.nlm.nih.gov/nu-ccore/-NC_000023.11?report=genbank &from=47078355&to=47093314) (Figure 2). All isoforms are biosynthesized by alternative splicing. Intact RGN is the translated product of exons 3 to 8 (299 amino acids). Isoform 1 (246 amino acids) is deleted the N-terminal region of RGN, since isoform 1 lacks exon 3 where the translation of RGN is initiated. Isoform 2 (227 amino acids) is a translated product in which exon 5 is deleted. Isoform 3 (183 amino acids) is a product in which both exon 5 and 6 have been deleted.

The expression of mRNA was confirmed in the liver of humans, rats, mice, bovines and chickens by northern blotting with an RGN cDNA probe (approximately 0.9 kb) [18]. In humans, two (or three) distinct mRNA isoforms of 1.7 and 1.5 kb were identified. The major RGN transcript in humans was the 1.7-kb one, since the band was significantly more intense than in the 1.5-kb isoform. Rats and mice also showed 1.7- and 1.5-kb transcripts. In contrast, mRNA in bovine liver gave double bands of 1.6 and 1.4 kb, and chicken liver had a single band about 1.5 kb in size.

A

```
                                                                                  80
RGN         MSSIKIECVL PENCRCGESP VWEEVSNSLL FVDIPAKKVC RWDSFTKQVQ RVTMDAPVSS VALRQSGGYV ATIGTKFCAL
isoform 1   ---------- ---------- ---------- ---------- ---------- ---MDAPVSS VALRQSGGYV ATIGTKFCAL
isoform 2   MSSIKIECVL PENCRCGESP VWEEVSNSLL FVDIPAKKVC RWDSFTKQVQ RVTMDAPVSS VALRQSGGYV ATIGTKFCAL
isoform 3   MSSIKIECVL PENCRCGESP VWEEVSNSLL FVDIPAKKVC RWDSFTKQVQ RVTMDAPVSS VALRQSGGYV ATIGTKFCAL

                                                                                  160
RGN         NWKEQSAVVL ATVDNDKKNN RFNDGKVDPA GRYFAGTMAE ETAPAVLERH QGALYSLFPD HHVKKYFDQV DISNGLDWSL
isoform 1   NWKEQSAVVL ATVDNDKKNN RFNDGKVDPA GRYFAGTMAE ETAPAVLERH QGALYSLFPD HHVKKYFDQV DISNGLDWSL
isoform 2   NWKEQSAVVL ATVDNDKKNN RFNDGKVDPA GRYFAA---- ---------- ---------- ---------- ----------
isoform 3   NWKEQSAVVL ATVDNDKKNN RFNDGKVDPA GRYFAG---- ---------- ---------- ---------- ----------

                                                                                  240
RGN         DHKIFYYIDS LSYSVDAFDY DLQTGQISNR RSVYKLEKEE QIPDGMCIDA EGKLWVACYN GGRVIRLDPV TGKRLQTVKL
isoform 1   DHKIFYYIDS LSYSVDAFDY DLQTGQISNR RSVYKLEKEE QIPDGMCIDA EGKLWVACYN GGRVIRLDPV TGKRLQTVKL
isoform 2   ---------- ---------- --------NR RSVYKLEKEE QIPDGMCIDA EGKLWVACYN GGRVIRLDPV TGKRLQTVKL
isoform 3   ---------- ---------- ---------- ---------- ---------- ---------- ---------- --KRLQTVKL

RGN         PVDKTTSCCF GGKNYSEMYV TCARDGMDPE GLLRQPEAGG IFKITGLGVK GIAPYSYAG                     299
isoform 1   PVDKTTSCCF GGKNYSEMYV TCARDGMDPE GLLRQPEAGG IFKITGLGVK GIAPYSYAG                     246
isoform 2   PVDKTTSCCF GGKNYSEMYV TCARDGMDPE GLLRQPEAGG IFKITGLGVK GIAPYSYAG                     227
isoform 3   PVDKTTSCCF GGKNYSEMYV TCARDGMDPE GLLRQPEAGG IFKITGLGVK GIAPYSYAG                     183
```

B

Figure 2. A comparison of RGN with three isoforms. A. Amino acid sequences of human RGN and three isoforms. Dashed lines show gaps inserted to supplement optional alignment between sequences. The blue- and green-stained amino acids indicate residues necessary for the structure involved in the formation of a cavity that binds to the substrate as gluconolactonase and for the coordination of the binding of a divalent cation, respectively. The numbers indicate the length of the amino acid sequences of RGN and its isoforms. B. Organization of the amino acid sequence of RGN and its isoforms. The line near the middle of isoforms 2 and 3 indicates that amino acid residues in this region have been deleted.

Murata and Yamaguchi identified a promoter region that binds to rat liver nuclear extract upstream of exon 2 according to an electrophoretic mobility shift assay (EMSA) [19]. They also reported that CAAT-box (5'-CCAAT-3'), AP-2 (5'-CCCCAGGC-3') and AP-4 (5'-CAGCTG-3') as *cis*-acting elements exist in this region [20]. Furthermore, nuclear extracts from the brain, spleen and heart specifically bind to the exon 2-derived fragment. The presence of multiple predicted promoter regions may be involved in the diversity of isoforms. Although a detailed analysis of the isoforms and tissue-specific expression of RGN has not been performed yet, clarifying this is important for understanding the physiological functions of RGN in various tissues.

Whether or not these three isoforms have the same function as the original RGN is unclear. In particular, isoforms 2 and 3 lack a continuous stretch of five amino acids (E120 – P124) that are important for binding to the substrate when this molecule acts as gluconolactonase [21]. Furthermore, isoform 3 lacks two of the three amino acids (E18, N154, D204) necessary to bind divalent cations such as Ca^{2+} and Zn^{2+} [21, 22]. The biological significance of isoforms will become apparent in future studies.

EVOLUTIONARY ASPECTS OF RGN

To clarify the genomic conservation of the RGN gene, genomic Southern hybridization was performed using RGN cDNA (approximately 0.9 kb) containing all 299 translated amino acids [18]. As a result, RGN genomic DNA was detected in humans, monkeys, rats, mice, dogs, bovines, rabbits and chickens, but not in yeast (*Saccharomyces cerevisiae*).

The homology is high when comparing the amino acid sequences of humans and other mammalian RGN (Figure 3). Indeed, it is particularly high in primates, being almost identical. Many mammalian species share the following two features of the amino acid sequence: (1) the 11 amino acids (R15, P35, P57, R101, E120 – P124, D265 and G266) needed to form a cavity that binds to the substrate when this molecule acts as gluconolactonase are conserved [21] and (2) the 3 amino acids (E18, N154, D204) in the molecule that modulate the binding of divalent cations, such as Ca^{2+} and Zn^{2+} [21, 22], are also conserved. Therefore, it is strongly suggested that RGN in other mammals serves the same function as human RGN.

	1 80
Human	MSSIKIECVL PENCRCGESP VWEEVSNSLL FVDIPAKKVC RWDSFTKQVQ RVTMDAPVSS VALRQSGGYV ATIGTKFCAL
Chimpanzee	MSSIKIECVL PENCRCGESP VWEEASNSLL FVDIPAKKVC RWDSFTKQVQ RVTMDAPVSS VALRQSGGYV ATIGTKFCAL
Gorilla	MSSIKIECVL PENCRCGESP VWEEASNSLL FVDIPAKKVC RWDSFTKQVQ RVTMDAPVSS VALRQSGGYV ATIGTKFCAL
Orangutan	MSSIKIECVL PENCQCGESP VWEEPAKKVC RWDSLTKQVQ RVTVDAPVSS VALHRSGDYV ATIGTKFCAL
Monkey	MSSIKIECVL PENCRCGESP VWEEASDSLL FVDIPAKKLC RWDSLTKQVQ RVTMDAPVSS VALRQSGGYV ATIGTKFCAL
Horse	MSSIKIECVL PENCQCGESP VWEEASSSLL FVDIPAKKVC RWDALSKQVQ RMTVDAPVGS VALRQSGGYV ATIGTKFCAL
Bovine	MSSIKIECVL RENCHCGESP VWEEASNSLL FVDIPAKKVC RWDSLSKQVQ RVTVDAPVSS VALRQSGGYV ATVGTKFCAL
Sheep	MSSMKIECVL RENCRCGESP VWEEASNSLL FVDIPAKKVC RWDSLSKQVQ RVTMDAPVSS VALRQSGGYV ATVGTKFCAL
Pig	MSSIKIECVL RENCHCGESP VWGEASNSLL FVDIPAKKVC RWNALSKQVQ RVTLDAPVSS VALRQAGGYV ATVGTKFCAL
Dog	MSSIKIECVL RENCRCGESP VWEEAANSLL FVDIPAKKVC RWDSLSKSVQ HVTVDAPISS VALRQGRGYV ATIGTKFCAL
Cat	MSSVKIECVL PENCRCGESP VWEEASDSLL FVDIPAKKVC RWDSLSKGVQ QVTVDAPVSS VALRQSGGYV ATIGTKFCAL
Rat	MSSIKIECVL RENYRCGESP VWEEASKCLL FVDIPSKTVC RWDSISNRVQ RVGVDAPVSS VALRQSGGYV ATIGTKFCAL
Mouse	MSSIKVECVL RENYRCGESP VWEEASQSLL FVDIPSKIIC RWDTVSNQVQ RVAVDAPVSS VALRQLGGYV ATIGTKFCAL
Rabbit	MSSIKIECVL PENCHCGESP VWEEASGSLL FVDIPGKKFC RWNPLTKAVQ RMTMDAPVTS VALRKSGGYV ATVGTKFCAL

	160
Human	NWKEQSAVVL ATVDNDKKNN RFNDGKVDPA GRYFAGTMAE ETAPAVLERH QGALYSLFPD HHVKKYFDQV DISNGLDWSL
Chimpanzee	NWKEQSAVVL ATVDNDKKNN RFNDGKVDPA GRYFAGTMAE ETAPAVLERH QGALYSLFPD HHVKKYFDQV DISNGLDWSL
Gorilla	NWKEQSAVVL ATVDNDKKNN RFNDGKVDPA GRYFAGTMAE ETAPAVLERH QGALYSLFPD HHVKKYFDQV DISNGLDWSL
Orangutan	NWKEQSAVVL ATVDNDKKNN RFNDGKVDPA GRYFAGTMAE ETAPAVLERH QGALYSLFPD HHVKKYFDQV DISNGLDWSL
Monkey	NWEEQSAVVL ATVDNDKKNN RFNDGKVDPA GRYFAGTMAE ETAPAVLERH QGSLYSLFPD HHVKKYFDQV DISNGLDWSL
Horse	NWEDQSVVVL AAVEKDKKNN RFNDGKVDPA GRYFAGTMAE ETAPAVTERH QGSLYALFPD HHVEKYFDQV DISNGLDWSL
Bovine	NWEDQSAVVL ATVDKEKKNN RFNDGKVDPA GRYFAGTMAE ETAPAVLERR QGSLYSLFPD HHVEKYFDQV DISNGLDWSM
Sheep	NWEDQSAVVL ATVDKDKKNN RFNDGKVDPA GRYFAGTMAE ETAPAVLERR QGSLYSLFPD HHVEKYFDQV DISNGLDWSM
Pig	NWEDESAVVL AAVDKDKKNN RFNDGKVDPA GRYFAGTMAE ETAPAVLERH QGSLYALFAD HHVEKYFDQV DISNGLDWSL
Dog	NWEDQSAVAL ATVDKDKKNN RFNDGKVDPA GRYFAGTMAE ETAPAVLERH QGSLYSLFPD GHVEKYFDQV DISNGLDWSL
Cat	NWEDQSVAVL ATVDKDKKNN RFNDGKVDPA GRYFAGTMAE ETAPAVLERH QGSLYSLFPD HHVEKYFDQV DISNGLDWSL
Rat	NWENQSVFIL AMVDEDKKNN RFNDGKVDPA GRYFAGTMAE ETAPAVLERH QGSLYSLFPD HSVVKYFDQV DISNGLDWSL
Mouse	NWENQSVFVL AMVDEDKKNN RFNDGKVDPA GRYFAGTMAE ETAPAVLERH QGSLYSLFPD HSVKKYFDQV DISNGLDWSL
Rabbit	NLEDQSVVAL ATVDKDKKNN RFNDGKVDPA GRYFAGTMAE ETAPAVLERH QGSLYALFPD HQVKKYFDQV DISNGLDWSL

	240
Human	DHKIFYYIDS LSYSVDAFDY DLQTGQISNR RSVYKLEKEE QIPDGMCIDA EGKLWVACYN GGRVIRLDPV TGKRLQTVKL
Chimpanzee	DHKIFYYIDS LSYSVDAFDY DLQTGQISNR RSVYKLEKEE QIPDGMCIDA EGKLWVACYN GGRVIRLDPV TGKRLQTVKL
Gorilla	DHKIFYYIDS LSYSVDAFDY DLQTGQISNR RSVYKLEKEE QIPDGMCIDA EGKLWVACYN GGRVIRLDPV TGKRLQTVKL
Orangutan	DHKIFYYIDS LSYSVDAFDY DLQTGQISNR RSVYKLEKEE QIPDGMCIDA EGKLWVACYN GGRVIRLDPV TGKRLHTVKL
Monkey	DHKIFYYIDS LSYSVDAFDY DLQTGQISNR RSVYKLEKEE QIPDGMCIDA EGKLWVACYN GGRVIRLDPV TGKRLQTVKL
Horse	DHKIFYYIDS LSYSVDAFDY DLLTGRISNR RSVYKLEKEE HIPDGMCIDT EGKLWVACYN GGRVIRLDPE TGKRLQTVKL
Bovine	DHKIFYYIDS LSYSVDAFDY DLQTGKISNR RSVYKLEKEE QIPDGMCIDV EGKLWVACYN GGRVIRLDPE TGKRLQTVKL
Sheep	DHKIFYYIDS LSYSVDAFDY DLQTGKISNR RSVYKLEKEE QIPDGMCIDV EGKLWVACYN GGRVIRLDPE TGKRLQTVKL
Pig	DHKIFYYIDS LSYSVDAFDY DLQTGKISNR RSIYKMEKEE HIPDGMCIDT EGKLWVACYN GGRVIRLDPE TGKRLQTVKL
Dog	DHKIFYYIDS LSYSVDAFDY DLQTGKISNR RSVYKLEKEE QIPDGMCIDT EGKLWVACYN GGRVIRLDPE TGKRLQTVKL
Cat	DHKIFYYIDS LSYSVDAFDY DLQTGKISNR RSVYKLEKEE QIPDGMCIDA EGKLWVACYN GGRVIRLDPE TGKRLQTVKL
Rat	DHKIFYYIDS LSYTVDAFDY DLPTGQISNR RTVYKMEKDE QIPDGMCIDV EGKLWVACYN GGRVIRLDPE TGKRLQTVKL
Mouse	DHKIFYYIDS LSYTVDAFDY DLQTGQISNR RIVYKMEKDE QIPDGMCIDA EGKLWVACYN GGRVIRLDPE TGKRLQTVKL
Rabbit	DHKIFYYIDS LAYSVDAFDY DLQTGQISNR RSIYKLEKEE QIPDGMCIDT EGKLWVACYN GGRVIRLDPE TGKRLQTVKL

	299	Homology % vs Human
Human	PVDKTTSCCF GGKNYSEMYV TCARDGMDPE GLLRQPEAGG IFKITGLGVK GIAPYSYAG	
Chimpanzee	PVDKTTSCCF GGKNYSEMYV TCAQDGMDPE GLLRQPEAGG IFKITGLGVK GIAPYSYAG	99.3
Gorilla	PVDKTTSCCF GGKNYSEMYV TCARDGMDPE GLLRQPEAGG IFKITGLGVK GIAPYSYAG	99.7
Orangutan	PVDKTTSCCF GGKNYSEMYV TCARDGMDPE GLLRQPEAGG IFKITGLGVK GIAPYSYAG	97.7
Monkey	PVDKTTSCCF GGKNYSEMYV TCARDGMDPE GLLRQPEAGG IFKITGLGVK GIAPYSYAG	98.0
Horse	PVDKTTSCCF GGKDYSEMYV TCARAGLDPE ALSRQPEAGG IFKITGLGVK GIPPYAYAG	90.0
Bovine	PVDKTTSCCF GGKDYSEMYV TCARDGLDSK GLLQQPEAGG IFKITGLGVK GIPPYPYTG	91.3
Sheep	PVDKTTSCCF GGKDYSEMYV TCARDGLDSK GLLQQPEAGG IFKITGLGVK GIPPYPYTG	91.6
Pig	PVDKTTSCCF GGKDYSEMYV TCARDGLDPQ GLLQQPEAGG IFKITGLGVK GLPPYPYAG	89.0
Dog	PVDKTTSCCF GGKDYSEMYV TCARDGMDPE SLLKQPEAGG IFKITGLGVK GIPPYPYAG	91.6
Cat	PVDKTTSCCF GGKDYSEMYV TCARDGMDAE KLLQQPQAGG IFKITGLGVK GIPPYPYAG	92.0
Rat	PVDKTTSCCF GGKDYSEMYV TCARDGMSAE GLLRQPDAGN IFKITGLGVK GIAPYSYAG	89.0
Mouse	PVDKTTSCCF GGKDYSEMYV TCARDGLNAE GLLRQPDAGN IFKITGLGVK GIAPYSYAG	88.6
Rabbit	PVDKTTSCCF GGKDYSEMYV TCARDGLDPD SLSRQPEAGG IFKITGLGVK GIPPYSYAG	89.3

Figure 3. A comparison of the amino acid sequences of human RGN and those of other mammals. The amino acid sequences of RGN in 14 species of mammals. Species names in red indicate primates. The blue- and green-stained amino acids indicate residues necessary for the structure involved in the formation of a cavity that binds to the substrate as gluconolactonase and for the coordination of the binding of a divalent cation, respectively. The yellow-stained amino acids are residues different from that of human RGN. The numbers indicate the percentage of homology to human sequences.

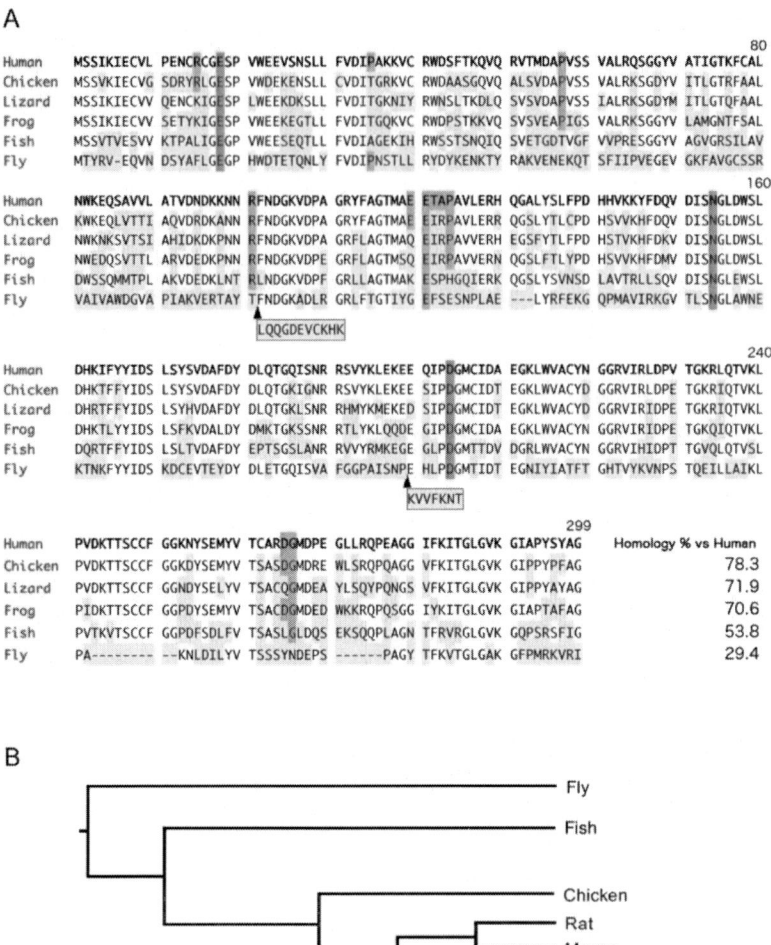

Figure 4. A comparison of the amino acid sequences of human RGN and those of other non-mammalian species. A. The amino acid sequences of six non-mammalian species of RGN. Dashed lines and the sequences in outside of row show gaps inserted to supplement optional alignment between sequences. The blue- and green-stained amino acids indicate residues necessary for the structure involved in the formation of a cavity that binds to the substrate as gluconolactonase and for the coordination of the binding of a divalent cation, respectively. The yellow-stained amino acids are residues different from that of human RGN. The numbers indicate the percentage of homology to human sequences. B. Phylogenetic tree of RGN amino acid sequences in six non-mammalian species. The phylogenetic tree was constructed with the neighbor-joining method using the CLUSTAL W software program [24].

However, in animals other than mammals, particularly in fish and flies, the two features of the amino acid sequence described above are scarcely conserved (Figure 4A); indeed, the homology of fish and fly sequences to the human amino acid sequence are 53.8% and 29.4%, respectively. The function of RGN may have been changed through the process of evolution. A phylogenetic tree analysis (Figure 4B) supports this notion.

STRUCTURAL FEATURES OF RGN

The most important function of RGN is the binding of calcium and its intracellular signal transduction. Many Ca^{2+}-binding proteins have an EF-hand motif as the common structural feature, which binds Ca^{2+} [23]. The S-100 protein has two EF-hands in a region with relatively high hydrophility. The most common EF-hand is composed of the helix-loop-helix domain. The prototype loop consists of 12 amino acids, 5 of which have a carboxyl (or a hydroxyl group) in their side chain, precisely spaced so as to coordinate the Ca^{2+}. An analysis of the structure of the EF-hand from the RGN sequence did not give the expected pattern of amino acids conforming to the typical EF-hand structure of a Ca^{2+}-binding site. Furthermore, the predicted amino acid sequence of RGN did not have significant homology with several other Ca^{2+}-binding proteins; indeed, RGN had low homology for calmodulin, calbindin-D28k and S-100β. Therefore, RGN is a novel protein that is completely different from other Ca^{2+} binding proteins in terms of its manner of binding with Ca^{2+}.

The location of three amino acid sequences of the helix, sheet and turn important for predicting the protein structure are shown in Figure 5A. Chakraborti and Bahnson determined the Ca^{2+}-binding site of RGN/SMP30 by analyzing the crystal structure in 2010 [22]. As shown in Figure 5B, they found that the RGN has a 6-bladed β-propeller fold that contains the binding site of a single metal ion with either a Ca^{2+} or a Zn^{2+} in the center. They confirmed that a mutation in the amino acid forming the Ca^{2+}-binding site significantly reduces the activity of RGN. Furthermore, Ca^{2+} had a significantly higher dissociation constant (Kd) than other divalent metal cations that were tested. Therefore, the authors concluded that the Ca^{2+}-bound form of RGN might be physiologically relevant for stressed cells with an elevated free calcium level.

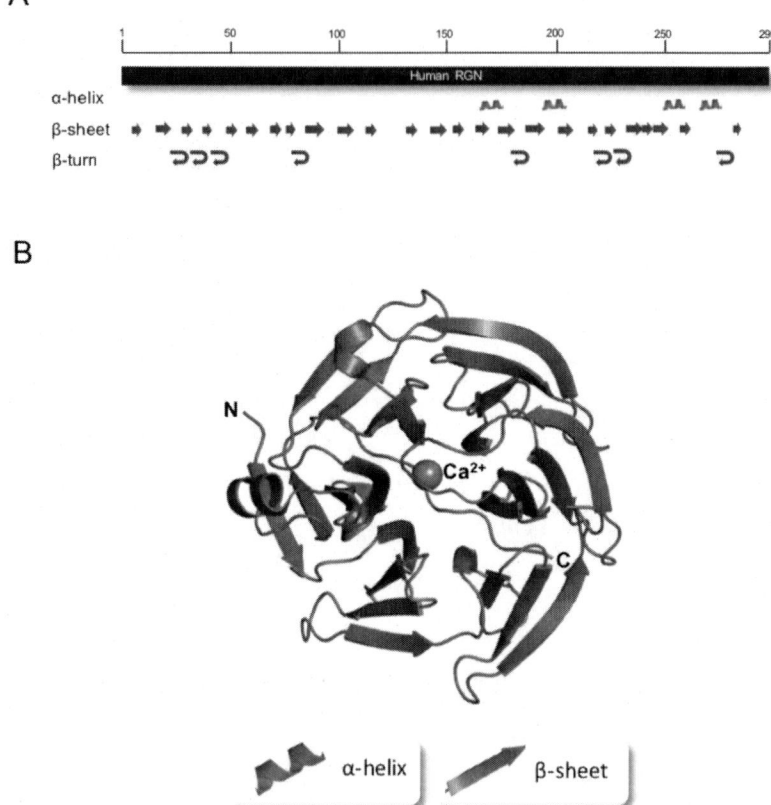

Figure 5. The structure of human RGN. A. The location of the three distinctive structures (α-helix, β-sheet and β-turn) in amino acid sequences that create the secondary structure of RGN. B. Crystal structure of human RGN with calcium binding (https://pubs.acs.org/doi/abs/10.1021/ bi9022297, https://www.ebi.ac.uk/pdbe/entry/pdb/ 3g4e). N and C represent the N-terminus and C-terminus of the protein, respectively. The Ca^{2+} located at the center of the structure is shown as a gray sphere.

However, Aizawa et al. reported the RGN/SMP30 structure in detail [21]. When this molecule acts as gluconolactonase (EC 3.1.1.17), R15, P35, P57, R101, E120 – P124, D265 and G266 are required for the formation of a cavity that binds to the substrate. In addition, three amino acids (E18, N154, D204) in the molecule coordinate the binding of divalent cations, such as Ca^{2+} and Zn^{2+} [21, 22]. These structural features indicate that RGN has activity as gluconolactonase and is a molecule that binds divalent cations.

CONCLUSION

Ca^{2+} plays an important role in various intra/extracellular events. At that time, the Ca^{2+}-binding protein regulates the kinetics of Ca^{2+} definitely. In this chapter, the gene and protein structure of RGN, a Ca^{2+}-binding protein that useful to study of Ca^{2+}-signaling have been described. These findings are very significant in considering RGN gene expression and function. In the future, RGN will be analyzed as a key molecule that is involved in various important functions beyond the framework of Ca^{2+}-signaling.

REFERENCES

[1] Yamaguchi M, Yamamoto T (1978) Purification of calcium binding substance from soluble fraction of normal rat liver. *Chem Pharm Bull* (Tokyo) 26:1915-1918.

[2] Yamaguchi M, Mori S (1988) Effect of Ca^{2+} and Zn^{2+} on 5'-nucleotidase activity in rat liver plasma membranes: hepatic calcium-binding protein (regucalcin) reverses the Ca^{2+} effect. *Chem Pharm Bull* (Tokyo) 36:321-325.

[3] Yamaguchi M (2000) Role of regucalcin in calcium signaling. *Life Sci* 66:1769-1780.

[4] Yamaguchi M, Sugii K (1981) Properties of calcium-binding protein isolated from the soluble fraction of normal rat liver. *Chem Pharm Bull* (Tokyo) 29:567-570.

[5] Shimokawa N, Yamaguchi M (1992) Calcium administration stimulates the expression of calcium-binding protein regucalcin mRNA in rat liver. *FEBS Lett* 305:151-154.

[6] Shimokawa N, Yamaguchi M (1993) Expression of hepatic calcium-binding protein regucalcin mRNA is mediated through Ca^{2+}/calmodulin in rat liver. *FEBS Lett* 316:79-84.

[7] Yamaguchi M (2014) Regucalcin as a potential biomarker for metabolic and neuronal diseases. *Mol Cell Biochem* 391:157-166. doi: 10.1007/s11010-014-1998-2.

[8] Yamaguchi M (2015) Involvement of regucalcin as a suppressor protein in human carcinogenesis: insight into the gene therapy. *J Cancer Res Clin Oncol* 141:1333-1341. doi: 10.1007/s00432-014-1831-z.

[9] Yamaguchi M (2015) The potential role of regucalcin in kidney cell regulation: Involvement in renal failure. *Int J Mol Med.* 36:1191-1199. doi: 10.3892/ijmm.2015.2343.

[10] Fujita T, Uchida K, Maruyama N (1992) Purification of senescence marker protein-30 (SMP30) and its androgen-independent decrease with age in the rat liver. *Biochim Biophys Acta* 1116:122-128.

[11] Fujita T, Shirasawa T, Uchida K, Maruyama N (1992) Isolation of cDNA clone encoding rat senescence marker protein-30 (SMP30) and its tissue distribution. *Biochim Biophys Acta* 1132:297-305.

[12] Shimokawa N, Yamaguchi M (1993) Molecular cloning and sequencing of the cDNA coding for a calcium-binding protein regucalcin from rat liver. *FEBS Lett* 327:251-255.

[13] Yamaguchi M, Isogai M (1993) Tissue concentration of calcium-binding protein regucalcin in rats by enzyme-linked immunoadsorbent assay. *Mol Cell Biochem* 122:65-68.

[14] Kozak M (1987) An analysis of 5'-noncoding sequences from 699 vertebrate messenger RNAs. *Nucleic Acids Res* 15:8125-8148.

[15] Shimokawa N, Matsuda Y, Yamaguchi M (1995) Genomic cloning and chromosomal assignment of rat regucalcin gene. *Mol Cell Biochem* 151:157-163.

[16] Yamaguchi M, Makino R, Shimokawa N (1996) The 5' end sequences and exon organization in rat regucalcin gene. *Mol Cell Biochem* 165:145-150.

[17] Breathnach R, Benoist C, O'Hare K, Gannon F, Chambon P (1978) Ovalbumin gene: evidence for a leader sequence in mRNA and DNA sequences at the exon-intron boundaries. *Proc Natl Acad Sci USA* 75:4853-4857.

[18] Shimokawa N, Isogai M, Yamaguchi M (1995) Specific species and tissue differences for the gene expression of calcium-binding protein regucalcin. *Mol Cell Biochem* 143:67-71.

[19] Murata T, Yamaguchi M (1998) Tissue-specific binding of nuclear factors to the 5'-flanking region of the rat gene for calcium-binding protein regucalcin. *Mol Cell Biochem* 178:305-310.

[20] Murata T, Yamaguchi M (1999) Promoter characterization of the rat gene for Ca^{2+}-binding protein regucalcin. Transcriptional regulation by signaling factors. *J Biol Chem* 274:1277-1285.

[21] Aizawa S, Senda M, Harada A, Maruyama N, Ishida T, Aigaki T, Ishigami A, Senda T (2013) Structural basis of the γ-lactone-ring formation in ascorbic acid biosynthesis by the senescence marker protein-30/gluconolactonase. *PLoS One* 8:e53706. doi: 10.1371/journal.pone.0053706.

[22] Chakraborti S, Bahnson BJ (2010) Crystal structure of human senescence marker protein 30: insights linking structural, enzymatic, and physiological functions. *Biochemistry* 49:3436-3444. doi: 10.1021/bi9022297.

[23] Taylor DA, Sack JS, Maune JF, Beckingham K, Quiocho FA (1991) Structure of a recombinant calmodulin from Drosophila melanogaster refined at 2.2-A resolution. *J Biol Chem* 266:21375-21380.

[24] Thompson JD, Higgins DG, Gibson TJ (1994) CLUSTAL W: improving the sensitivity of progressive multiple sequence alignment through sequence weighting, position-specific gap penalties and weight matrix choice. *Nucleic Acids Res* 22: 4673-4680. doi: 10.1093/nar/22.22.4673.

In: Regucalcin
Editor: Masayoshi Yamaguchi

ISBN: 978-1-53616-172-4
© 2019 Nova Science Publishers, Inc.

Chapter 2

THE HORIZONS OF REGUCALCIN IN MALE REPRODUCTION

Sara Correia[1], Harikrishna Pillai[2], Ana Manuela S. Silva[1], Vanessa M. Santos[1] and Silvia Socorro[1,*]
[1]CICS-UBI - Health Sciences Research Centre, University of Beira Interior, Covilhã, Portugal
[2]Animal Husbandry Department, Government of Kerala, Kerala, India

ABSTRACT

Regucalcin (RGN) is a highly conserved calcium (Ca^{2+})-binding protein, present throughout the evolution line, from prokaryotes to eukaryotes. It is highly expressed in the liver and kidney of mammalian vertebrates, but its broad expression in a panoply of other tissues has been reported. Over the years, the RGN role in the regulation of several biological processes apart from Ca^{2+} homeostasis also was being disclosed. RGN was shown to be a multifunctional protein involved in the regulation of intracellular signaling pathways, oxidative stress, cell

---

[*] Corresponding Author's E-mail: ssocorro@fcsaude.ubi.pt.

proliferation, apoptosis, and also energetic metabolism. All these are crucial processes for the successful production of male gametes, which propelled researchers investigating the RGN role in male reproduction. The present chapter systematically describes the existent literature on the actions of RGN supporting spermatogenesis and sperm function, and also discusses the recent experimental advances that sustain the usefulness of this protein in reproductive technology.

Keywords: regucalcin, reproduction, male fertility, spermatogenesis, sperm capacitation, cryopreservation, sex-steroid target gene, calcium, oxidative stress, apoptosis

INTRODUCTION

Regucalcin (RGN) is a calcium (Ca^{2+})-binding protein highly conserved throughout evolution, being present from prokaryotes to the different levels of complexity of eukaryotes [1-10]. Its role in the regulation of intracellular Ca^{2+} homeostasis has been shown to occur through the modulation Ca^{2+}-pumps activity at plasma membrane, endoplasmic reticulum and mitochondria [11, 12]. However, RGN ability to bind other divalent cations also has been reported [13, 14].

The full-length RGN protein has 299 amino acid residues and an approximate molecular weight of 33-34 kDa, which together with a characteristic downregulation with aging led some authors naming it Senescence Marker Protein-30 (SMP-30) [5, 13, 15].

Although being highly expressed in the liver and kidney cortex, RGN expression has been detected in a broad range of tissues and body fluids of several vertebrate and invertebrate species [16]. Since 2008, this has encompassed the characterization of RGN expression in the male reproductive tract, including, the gonads, accessory glands, excurrent ducts fluids and spermatozoa [17-21].

Also, a panoply of factors has been shown to regulate RGN tissue expression levels, namely, Ca^{2+}, glucose levels, oxidative stress, insulin, thyroid and parathyroid hormones, and steroid hormones [16, 22].

Moreover, RGN expression is modulated in response to different physiological conditions and diseases states [23-28].

The last decade has witnessed the emergence of RGN as a multifunctional protein regulating distinct biological processes besides the maintenance of Ca^{2+} homeostasis. For example, RGN has been shown to regulate intracellular signaling pathways by influencing the activity of kinases, phosphatases, phosphodiesterase, and proteases [12, 29]. Several reports have been presented concerning the antioxidant properties of RGN, which described its influence diminishing oxidative stress, increasing the antioxidant defense, and protecting from age deterioration of cell function [16, 30]. Interestingly, accumulating evidence has placed RGN as an important regulator of tissues homeostasis by its dual action controlling proliferation and apoptosis [30-33]. Therefore, its role as a cytoprotector and anti-tumor protein has been proposed [22, 34, 35]. More recently, the properties of RGN modulating cell metabolism started being disclosed and have been related to the control of glycolytic metabolism and lipid handling [36, 37].

The strict control of oxidative stress, the delicate balance between cell death and proliferation, the protection from damaging exogenous factors, as well, as the proper metabolic support to ensure successful germ cell development, are determinant issues for successful spermatogenesis and, thus, male fertility. Therefore, in the last years, the study of RGN in the context of male reproduction has deserved the attention of some research groups.

THE EVOLUTIONARY HISTORY OF RGN

The RGN protein has been identified in a large number of species, being present in invertebrates, mammalian and nonmammalian vertebrates, as well as in fungi and bacteria [1-10]. To disclose the peculiarities of RGN protein structure and its functional domains, researchers used protein sequence alignment tools and determined the amino acid identities among RGN's sequences from several species [13, 16, 38]. The RGN gene is highly conserved from eukaryotes to prokaryotes, and 18% of residues in RGN protein are conserved in all species [38]. Overall, the amino acid sequence of the human RGN protein (299 residues long, 34 kDa) shows 98% of similarity with other primates, 93–96% with other mammalian species and 79–85% with non-mammalian vertebrates (reviewed by [16]). Moreover, human RGN shows 43 to 47% of similarity with invertebrates, bacteria and fungi [16], which is amazing considering the extremely high difference of organisms' complexity being compared. These findings represent irrefutable evidence about the importance of RGN since the very beginning of life and reinforced the idea that it should play a relevant basic cellular function well-conserved throughout evolution [16].

Motif analysis showed that RGN do not has the typical EF-hand Ca^{2+}-binding sequence but the crystal structure of the human protein, solved by X-ray diffraction [14], UV difference, fluorescent emission and circular dichroism studies [39, 40] confirmed binding of Ca^{2+} ions. Nevertheless, X-ray diffraction together with mutagenesis analysis, also showed that RGN could also bind other divalent cations such as zinc, manganese and magnesium [13, 14]. In addition, RGN was shown to have enzymatic activity functioning as a gluconolactonase, an enzyme that catalyzes the penultimate step in the biosynthesis of L-ascorbic acid [41].

RGN is a protein with a wide tissue distribution including the male reproductive tract (discussed in sections 4 and 5). Also, it has a vast distribution within the distinct cell compartments, being detected in the cytoplasm [17, 18, 42-47], mitochondrial fractions [47] and peri-nuclear space, as well as, in the nucleus [17, 18, 42-45, 47-53]. Although several reports have shown the presence of RGN in the nucleus, no study has demonstrated the existence of a functional nuclear localization signal (NLS) in RGN protein sequence. Nevertheless, analysis of the first 60 N-terminal amino acid residues of mammalian RGN proteins (Figure 1) allowed to identify an importin α-dependent putative NLS. Noteworthy, ~72% (23 of 32) of the residues of the putative NLS (Figure 1) are 100% conserved in all species analyzed.

The RGN gene is localized in the q11.1-12 and p11.3-q11.2 segments of the rat and human X chromosomes, respectively [54, 55]. Besides the full-length mRNA encoding the 299 residue RGN protein, two alternatively spliced mRNA variants, originated by an exon skipping mechanism, have been described. RGNΔ4 transcript has a deletion of exon 4, and the RGNΔ4,5 transcript lacks exons 4 and 5, which would likely correspond to the RGN lower molecular protein variants that have been identified (28 and 24 kDa) [47]. These mRNA transcripts have been identified in several human tissues, namely, breast, prostate and testis [38, 51], which suggests that they are functionally relevant, but their biological role remains to be identified.

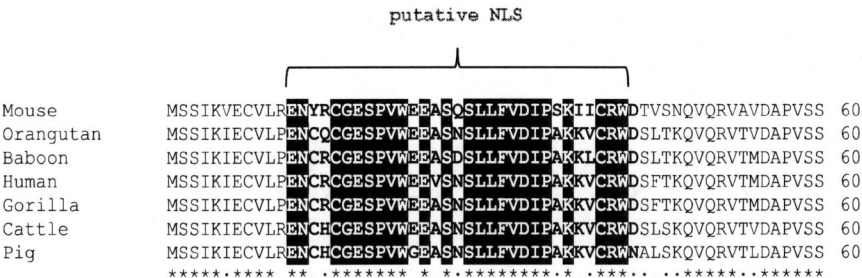

Figure 1. Multiple sequence alignment of the first 60 N-terminal amino acid residues of mammalian RGN proteins showing a putative importin α-dependent nuclear localization signal (NLS) predicted using the cNLS Mapper [62]. The putative NLS was obtained with a score of 4 that is typical of a protein localized to both the nucleus and the cytoplasm. Alignment was produced with Clustal Omega alignment tool [63]. Black shading indicates residues of the putative NLS 100% conserved in all species. Latin names and Genbank accession numbers for corresponding sequences were as follows: mouse, *Mus musculus* (NP_033086.1); orangutan, *Pongo abelii* (NP_001127502.1); baboon, *Papio anubis* (XP_003917675.1); human, *Homo sapiens* (NP_004674.1); gorilla, *Gorilla gorilla* (XP_018874621); cattle, *Bos taurus* (NP_776382.1); pig, *Sus scrofa* (NP_001070688.1).

The fact that the RGN protein is encoded by an X-linked gene is very interesting when envisaging its liaison with male reproduction. Y chromosome was initially thought as playing the major role in the regulation of male fertility. However, the enthusiastic efforts of scientific community allowed to discover that several genes located on X chromosome can modulate fertility [56]. This is, for example, the emblematic case of androgen receptor [57-59]. In fact, over the years it has been unrevealed that the maternal X chromosome is enriched in genes with high testicular expression, which play important roles in the regulation of spermatogenesis [60, 61]. Genetic studies have shown that several X-linked genes are implicated in the control of meiosis, playing critical functions in male germ cell maturation [61]. In fact, the X chromosome has been considered of particular interest in the study of human infertility attributed to the male factor [61]. Since males possess only one X chromosome, mutations in the single-copy of an X-linked gene would not be camouflaged by the normal allele and, therefore, would manifest. For this reason, X-linked genes, such as the case of RGN, would be especially important in male fertility and could give useful information into the clinical practice.

EVIDENCE OF RGN AS A SEX STEROID TARGET GENE

Hormone regulation and the complex interaction between hypothalamic, pituitary and gonadal hormones are crucial for mammalian reproductive function [64]. The steroid hormones produced by the gonads are normally referred to as sex steroid hormones because of their primary role in sex differentiation, gonadal function and germ cell development [64].

Androgens are recognized as the main regulators of spermatogenesis with their actions being fundamental for germ cells survival, maturation and sperm production [65-67] whereas also exerting negative feedback on the hypothalamus and pituitary regulating the spermatogenic output [68]. Though estrogens' roles in spermatogenesis are not so clear, it is widely known that both somatic and germ cells express nuclear and membrane estrogens receptors [69, 70], and that testicular cells are capable of synthesizing estrogens [71]. Moreover, a physiological role for estrogens in male fertility has been proposed as regulators of germ cell survival and apoptosis [71, 72].

Sex steroid hormones, androgens and estrogens, exert their biological actions by interaction with androgen and estrogen receptors, respectively.

This class of hormone receptors belong to the superfamily of nuclear transcription factors regulating gene expression network in cells and tissues dependently on environmental conditions and developmental stage [73, 74]. Since the emergence of modern molecular biology techniques in the 80s-90s years, the structure and function of steroid receptors, as well as, their molecular targets and tissue expression patterns, have been characterized (e.g., [69, 71, 75, 76]). In the reproduction field, the identification of new steroid target genes helped to clarify the molecular basis of spermatogenesis and the relevance of specific players for male fertility (e.g., [18, 71, 72, 77, 78]). With this rationale, several studies aimed to investigate the role of steroid hormones regulating RGN expression levels in reproductive tissues. First evidence showed that RGN is a target of estrogens. Subcutaneous administration of 17β-estradiol (E_2) augmented RGN mRNA expression in the rat liver [79], whereas producing the opposite effect in the kidney [80], mammary gland and prostate [17]. Also, in bovine bulbourethral glands and prostate, E_2 administration lead to decreased RGN expression [81]. Interestingly, E_2-treated MCF-7 breast cancer cells displayed up-regulated expression of RGN for 6 and 12 h of stimulation, but longer exposure times lead to a diminished expression of RGN [51], in accordance with the results in the rat mammary gland *in vivo* after 7 days of treatment [17].

The action of estrogens regulating RGN expression in the testis was assessed using the *ex vivo* culture of rat seminiferous tubules. E_2-treatment (100 nM) for 24 hours augmented RGN expression in intratubular testicular cells, which was suggested to be a response mechanism to counteract the enhanced apoptosis driven by estrogens [82]. Contrastingly, a study in bovines demonstrated that *in vivo* treatment (190 mg E_2/animal/day) caused down-regulation of RGN expression in the testis [81].

Concerning the regulation by androgens, orchidectomized rats displayed reduced RGN expression in the kidney that was increased after testosterone replacement [83]. However, regarding reproductive tissues, orchidectomy increased RGN expression in rat prostate that was down-regulated after 5α-dihydrotestosterone (DHT) administration [32].

Accordingly, treatment of LNCaP prostate cancer cells with DHT diminished RGN expression, an effect that seemed to be mediated by the androgen receptor involving *de novo* protein synthesis [51]. The androgen receptor was also implicated on the effect of DHT (10^{-7} M) up-regulating RGN expression in rat seminiferous tubules cultured *ex vivo* [18]. This was the first evidence confirming RGN as an androgen target gene. However, it was not followed by *in vivo* findings. In bovines, testosterone administration (1.050 g/animal/day) decreased RGN expression in the testis [81].

Overall, the existent literature has established RGN both as an estrogen- and androgen-target gene. However, sex steroid hormones actions modulating RGN expression levels seem to be tissue- and species-specific, and time-dependent, which might be due to specificities in the interaction with the DNA or differential recruitment of estrogens and androgen receptors co-regulators. Moreover, changes in RGN expression in response to steroids were suggested to be related to the suppression of germ cells apoptosis with RGN mediating these hormones' functions as survival factors in spermatogenesis.

Androgens and estrogens have been proposed to increase intracellular Ca^{2+} levels in several cell types [84-88]. Thus, considering the influence of RGN over Ca^{2+} handling [11, 21] and the importance of this ion in spermatogenesis and sperm function it would be relevant to explore the relationship between sex steroid hormones, RGN and intracellular Ca^{2+} levels in reproductive tissues.

EXPRESSION PATTERN OF RGN IN MALE REPRODUCTIVE TRACT AND SPERMATOZOA

The first indication about the association of RGN with reproduction came in 2004 when the RGN gene was shown to be expressed in the bovine ovarian follicle and reported to be associated with follicular growth and dominance. An 8-fold higher expression level was described in dominant follicles compared to that of small follicles [89], which was the very early indication of the supportive role of RGN for germ cells.

Table 1. RGN expression in tissues and cells of male reproductive tract

Tissue	Species	Localization	Biological form	Reference
Testis	Human	Cytoplasm and some nuclei of Leydig cells, Sertoli cells and germ cells	mRNA and Protein	[18]
	Bovine	Cytoplasm of Leydig cells and weak staining of nuclei of some spermatogonia	mRNA and Protein	[81]
	Buffalo	Cytoplasm and nucleus of Sertoli cells and germ cells	mRNA and Protein	[20]
	Rat	Cytoplasm and some nuclei of Leydig cells, Sertoli cells and germ cells	mRNA and Protein	[18]
	Mice	-	mRNA	[90]
Seminal vesicles	Buffalo	Cytoplasm and nuclei of glandular epithelial cells	mRNA and Protein	[20]
	Rat	Epithelium	mRNA and Protein	[18]
Epididymis	Buffalo	Cytoplasm and nuclei of glandular epithelial cells, luminal secretion and interstitial tissue	mRNA and Protein	[20]
	Rat	Epithelium, smooth muscles and connective tissue	mRNA and Protein	[18]
Prostate	Bovine	Cytoplasm of glandular epithelial cells		[81]
	Buffalo	Cytoplasm and nuclei of glandular epithelial cells	mRNA and Protein	[20]
	Rat	Epithelium	mRNA and Protein	[18]
Bulbo-urethral glands	Bovine	Nuclei of glandular cells	mRNA and Protein	[81]
	Buffalo	Cytoplasm and nuclei of glandular epithelial cells	mRNA and Protein	[20]
Ampulla of vas deferens	Buffalo	Nuclei of glandular epithelium and interstitial tissue	mRNA and Protein	[20]
Spermatozoa	Buffalo	Acrosome	mRNA and Protein	[19]

Later, *RGN* mRNA was identified in mice testis [90], and the first evidence of RGN protein expression in a broad range of male reproductive organs was reported for rat and human by Sílvia Socorro group in 2011 [18]. These findings were followed by the characterization of RGN expression in the testis, epididymis, prostate and seminal vesicles of other mammalian species, namely, buffalo and bovines. Table 1 summarizes the outcomes from the main studies reporting RGN expression in male reproductive organs and spermatozoa. In the testis, RGN was shown to be localized in the endocrine Leydig cells, in the intratubular somatic cells, the Sertoli cells, as well as in the germline [18].

Pillai et al. [19] were pioneer describing the presence of RGN in spermatozoa by localizing the protein to the acrosome region of buffalo sperm. Acrosome is the major Ca^{2+}-storage organelle of spermatozoa, and the localization of RGN in the acrosomal region propelled researches to focus on finding out this protein role in Ca^{2+}-related functions like capacitation, acrosomal reaction and membrane fusion (to be detailed in section 7). Apart from the three reported isoforms of RGN (34 kDa, 28 kDa and 24 kDa) [47], two new isoforms of 48 kDa and 44 kDa were reported in sperm [19]. The same study also showed the association of RGN to spermatozoa membrane and its relocation from the cytosol to the acrosomal region during maturation, i.e., from testicular spermatozoa to ejaculated spermatozoa.

Moreover, RGN has been identified as a secreted protein being detected in several fluids of male reproductive tract, namely, in seminiferous tubules, epididymis, and seminal vesicles fluids [20, 21, 83].

RGN Expression in Distinct Phenotypes of Human Spermatogenesis

As a clue to understand the liaison of RGN to male reproduction, its expression pattern in human testis with normal (Table 1) and abnormal phenotypes of spermatogenesis was investigated. RGN expression levels were analyzed in cases of hypospermatogenesis (HP) and Sertoli cell-only syndrome in comparison with those of testis with conserved spermatogenesis. Increased expression of RGN was found in the testis of men with HP comparatively with the cases of obstructive azoospermia with conserved spermatogenesis and Sertoli cell-only syndrome [38].

The aetiology of HP is complex and not entirely known, but it has been established that the development and maintenance of successful spermatogenesis depend on a delicate balance between germ cell proliferation and apoptosis [91]. Indeed, augmented rates of apoptosis and altered expression patterns of a panoply of cell death regulators have been detected in the testes of subfertile and infertile men [92-94]. Moreover, accelerated apoptosis has been implicated as a cause for the decreased number of spermatogonia in HP cases, rather than proliferative dysfunction in the mitotic phase [93].

Presently, no definitive conclusion can be drawn for the increased expression levels of RGN in HP cases. However, it is liable to speculate that RGN may be acting as a protective molecule counteracting the augmented rates of apoptosis associated with disrupted spermatogenesis. A dual role for RGN controlling cell cycle and apoptosis has been proposed [16, 22], with RGN expression being increased in response to the induction of apoptosis by chemical or physical agents [78, 95, 96]. Thus, the equilibrium of RGN expression levels would be determinant for the maintenance of the appropriate germ cell lineage number and, consequently, for successful spermatogenesis. The RGN' roles modulating apoptosis and as a protective molecule for male germ cells have been investigated in recent years and will be discussed further in the following sections.

RGN'S INFLUENCE IN MALE REPRODUCTION: WHAT WE HAVE LEARNED FROM THE TRANSGENIC RAT MODEL

After the description of Ca^{2+}-binding protein RGN in several tissues of male reproductive tract, including testis [18], and its identification as an estrogen- and androgen-target gene [17, 18, 51], the potential role of this protein in testicular function has been exploited. A substantial body of evidence has been produced using a transgenic rat model overexpressing RGN (Tg-RGN), which was originally generated by Yamaguchi M [97] by means of oocyte transgene pronuclear injection. These animals are fertile and their spermatogenic output was first investigated by Correia et al. [21]. Quantitative and qualitative sperm parameters, as well as the morphology and function of epididymis, were compared between Tg-RGN and their wild-type (Wt) littermates. Although able to breed, Tg-RGN animals displayed lower sperm counts and reduced sperm motility (Table 2). However, these features seem to be counterbalanced by the higher viability and diminished incidence of morphological defects (Table 2) found in the epididymal sperm of Tg-RGN rats [21].

Table 2. Sperm parameters in the Tg-RGN rats in comparison with their Wt littermates

Sperm parameters*	Tg-RGN	Wt	p-value
Sperm counts	$1.28 \times 10^8 \pm 9.24 \times 10^6$	$1.72 \times 10^8 \pm 1.57 \times 10^7$	$P < 0.05$
Motility	$47.88\% \pm 3.67$	$64.60\% \pm 5.66$	$P < 0.05$
Viability	$38.75\% \pm 2.36$	$28.00\% \pm 3.84$	$P < 0.05$
Normal morphology	$74.13\% \pm 3.74$	$57.58\% \pm 1.76$	$P < 0.01$
Tail defects	$18.60\% \pm 2.60$	$36.18\% \pm 2.04$	$P < 0.001$

*- Published in [21]; Values represent mean ± SEM.

It is through the passage in the epididymis, specifically in the order *caput*, *corpus* and *cauda*, and before being released in *vas deferens*, that sperm acquire the ability to move progressively and fertilize an oocyte [98]. The unique microenvironment of the epididymal lumen is mainly maintained by the secretory activity of the *caput* region [99]. The *corpus*' secretory activity is lower, playing a role in late sperm maturation events whereas *cauda* essentially stores the functionally mature sperm [100]. RGN expression was two-fold higher in the *corpus* relatively to *caput* and *cauda* regions, which emphasizes its importance for sperm maturation.

Interestingly, an altered morphology of *caput* was found in the epididymis of Tg-RGN rats. Although there were no differences in the tubule area, boundwidth, boundheight, and perimeter between Wt and Tg-

RGN groups, the epithelial cell height of this region was significantly decreased in Tg-RGN [21]. The altered morphology indicated important alterations in the reabsorptive/secretory activity of this epididymis region that could have an impact in sperm parameters. Indeed, diminished capacity of Ca^{2+} influx was detected in the epididymis of Tg-RGN rats, which suggested that Ca^{2+} concentrations are augmented in the epididymal fluid and that might be responsible for the reduced sperm motility observed in these animals [21].

Moreover, epididymal tissues of Tg-RGN animals showed a higher antioxidant potential compared with Wt [21]. Oxidative damage is one of the main factors leading to male germ cells death and increased incidence of defects [101]. It is also known that RGN has been linked with the decreased generation of reactive oxygen species and increased activity of antioxidant defense systems in several cell types, including liver, lung, heart and brain cells [29, 102-105]. This led the authors to assume that the higher sperm viability and the diminished incidence of morphological defects exhibited by Tg-RGN animals (Table 2) may be a consequence of the RGN' role protecting sperm from oxidative stress.

In this way, the antioxidant role of RGN in testicular cells was deepened in further research by exposing seminiferous tubules of Tg-RGN rats and controls to pro-oxidant stimuli, namely, tert-butyl hydroperoxide and cadmium chloride [78]. As hypothesized, Tg-RGN animals displayed increased protection against oxidative damage, exhibiting lower levels of oxidative stress demonstrated by diminished lipid peroxidation levels [78]. Also, antioxidant defenses, like glutathione S-transferase, were augmented in the Tg-RGN when exposed to oxidant conditions [78]. As expected, the lower oxidative damage in the seminiferous tubules of Tg-RGN rats was translated in reduced rates of apoptotic cell death [78].

Germ cells are highly sensitive to endogenous and exogenous damaging factors and, as discussed previously in section 5, increased apoptosis has been indicated as a cause of male infertility. The anti-apoptotic effect of RGN has been suggested in different cell lines and *in vivo* models involving the activation of multiple pathways and molecular targets, such as Akt, p53, Fas, transforming growth factor-β, tumor necrosis factor-α, caspases and several Bcl-2 family members (reviewed in [16, 22, 31]. The overexpression of RGN was shown to suppress thapsigargin- and actinomycin D-induced apoptosis in rat seminiferous tubules cultured *ex vivo* by modulating the expression and activity of key apoptotic and antiapoptotic factors [95]. Reduced expression and activity of the executioner of apoptosis caspase-3 were observed together with increased expression of p53 and Bcl-2 [95].

In vivo findings also demonstrated the protective role of RGN over apoptosis of the germ line. Tg-RGN animals shown to be resistant to radiation-induced testicular damage displaying lower rates of apoptosis after irradiation (reduced activity of caspase-3, lower levels of caspase-8, and increased Bcl-2/Bax ratio) [96]. Moreover, suppressed apoptosis was concomitant with higher testis volume, augmented sperm viability and motility, as well as a higher frequency of normal sperm morphology and diminished incidence of head-defects relatively to Wt counterparts.

Sertoli cells are the somatic cells responsible for providing the germ cells with physical support, as well as, with the adequate supply of growth factors and nutrients, namely, lactate that has been indicated as the preferred energy source for germ cells [106]. Very recently, the metabolic status of primary Sertoli cells cultures obtained from the testis of Tg-RGN rats was characterized. These cells, though consuming less glucose, produced high levels of lactate, and displayed increased expression of alanine transaminase, and augmented glutamine consumption indicating high plasticity of metabolic routes in response to RGN overexpression [107]. Moreover, the lactate produced seems to be consumed by the germ cells with a consequent diminution of apoptotic rate.

In sum, the Tg-RGN model contributed to establish the basis of the RGN' role in male reproduction. Also, it allowed a better understanding of the molecular regulation of spermatogenesis and sperm function opening new avenues of research, as well as, novel perspectives for the development of infertility treatments and contraceptive methods.

PROTECTIVE ROLES OF RGN FOR THE FERTILIZATION CAPACITY OF MAMMALIAN SPERMATOZOA

Anti-Oxidant Effect

Mammalian spermatozoa are highly susceptible to oxidative damage due to their higher content of polyunsaturated fatty acids. Therefore, the presence of antioxidants in the seminal fluid is very important to maintain the highly sensitive redox equilibrium of spermatozoa and their functionality. Superoxide dismutase, catalase and glutathione peroxidase are the major antioxidant enzymes present in semen playing a relevant role to reduce oxidative stress [108].

RGN is broadly present in male reproductive tract tissues (Table 1) and secretions till ejaculation [18, 20, 21]. After ejaculation RGN is removed from seminal plasma whereas spermatozoa maintain RGN expression localized at the acrosomal region [19, 20]. The antioxidant properties of RGN have been demonstrated by many studies in different cell lines and *in vivo* models [109-111]. RGN has been shown to reduce intracellular levels of oxidative stress through modulation of enzymes involved in the generation of free radicals, as well as, in the antioxidant defense [29, 102, 103, 112]. It has been demonstrated that superoxide dismutase activity was enhanced in normal rat liver and heart in the presence of exogenous RGN [29, 103] and the glutathione peroxidase activity was reduced in animals without RGN [109, 110]. Also, RGN overexpression *in vivo* was capable of counteracting aging-associated changes in rat prostate, maintaining low levels of oxidative stress, reducing lipid peroxidation and sustaining high activity levels of glutathione-S-transferase [30].

Furthermore, RGN has been identified as a gluconolactonase [41], an enzyme involved in the penultimate step of L-ascorbic acid synthesis in the liver. Ascorbic acid, a cofactor in metal-dependent oxygenases [113], has been indicated as an important antioxidant in semen that protects spermatozoa during cryopreservation [114]. As discussed earlier, a higher antioxidant potential was reported in the epididymis of Tg-RGN rats, which was linked to the higher sperm viability, higher percentage of normal morphology and diminished incidence of tail defects compared to Wt counterparts [21]. Moreover, the testis of Tg-RGN animals was shown to display increased antioxidant capacity and enhanced activity of glutathione-S-transferase [78]. Future research efforts are warranted to disclose the role of RGN as a gluconolactonase and modulating the activity of superoxide dismutase and glutathione peroxidase in the testis and epididymis, as well as, in sperm storage and cryopreservation.

Anti-Capacitatory Effect

The capacitation process represents the set of biochemical, structural and physiological changes that sperm undergo in the female reproductive tract rendering them able to fertilize an oocyte (reviewed recently in [115]). Seminal plasma and its components help to prevent the premature onset of sperm capacitation till ejaculation. Several decapacitation factors like phosphatidylcholine-binding proteins [116], semenogelin [117], cholesterol [118] and zinc [119] are present in seminal plasma. Moreover, the capacitation-associated tyrosine phosphorylation of sperm proteins is also shown to be inhibited by seminal plasma [120].

The decapacitation factors should be removed after ejaculation for the occurrence of capacitation and the subsequent acrosomal reaction, which ultimately allows spermatozoa to fertilize the ovum. The proteolytic enzymes released upon acrosome reaction enable the fusion of sperm and oocyte membranes [121]. RGN is present in the testis, sperm and throughout male excurrent ducts, including epididymal and vesicular fluids, and disappearing in the ejaculated seminal plasma. These findings indicated the potential anti-capacitatory role of RGN, which was investigated by Pillai et al. [20]. Co-incubation of recombinant RGN with buffalo seminal plasma confirmed the degradation of protein in the presence of seminal components [20]. Another evidence for the anti-capacitatory property of RGN is that the epididymal spermatozoa of Tg-

RGN showed a reduction in motility compared to their Wt counterparts [21].

The suppressive effect of RGN on *in vivo* capacitation was further studied in buffalo using the fluorescent cholortetracyclin assay to detect capacitated spermatozoa [20]. Addition of recombinant RGN to capacitating media significantly reduced the percentage of capacitated spermatozoa compared to the untreated group. Overall, gathered information indicating the anti-capacitory role of RGN raises the curiosity about its action in human sperm and the benefits of this protein in assisted reproduction techniques.

Cryoprotective Role of RGN in Spermatozoa

Cryopreservation of semen and artificial insemination are important procedures that allowed significant benefits to the livestock industry [122], as well as, introduced important advances in human reproduction technology [123]. Cryopreservation of human sperm started in the 1960s, and over the years it has been noted that the fertility potential of cryopreserved mammalian spermatozoa is lower than that of fresh sperm [124]. Cryopreservation induces extensive biophysical and biochemical changes in the membrane of spermatozoa that ultimately decrease their fertility potential [125]. Factors as sudden temperature changes, ice formation and osmotic stress have been proposed as the main reasons for poor sperm quality after cryopreservation and thawing [123]. In addition, the procedures of cryopreservation induce premature capacitation of spermatozoa, which is known as cryo-capacitation [122]. These alterations may not affect motility but reduce viability, the ability to interact with the female reproductive tract and sperm fertility. For example, 8 times more cryopreserved bovine spermatozoa were required to achieve equivalent fertilization rates *in vivo* compared to fresh sperm (reviewed in [122]).

A cytoprotective role has been proposed for RGN by its actions counteracting deregulated proliferation and apoptosis, as well as, minimizing Ca^{2+}-related stress and oxidative damage [16, 112].

Noteworthy, deregulation of Ca^{2+} levels and abnormal levels of reactive oxidant species are the main cyto-damaging factors during sperm cryopreservation. Most importantly, RGN was reported as a putative cold tolerance gene in *Drosophila montana* [126], which supported the advantageous role of this protein in sperm cryopreservation. Indeed, the cryoprotective role of recombinant RGN in buffalo spermatozoa was recently reported. Supplementation of buffalo semen extenders with 1 µM of recombinant RGN during freezing resulted in significant increases in the post-thaw progressive motility, acrosome integrity, and zona pellucida binding of spermatozoa compared to control conditions without RGN [127]. The ability of RGN to counteract the excessive production of reactive oxygen species and Ca^{2+}stress may be accountable to its cryoprotective effect in spermatozoa.

CONCLUSION

Reports of the anti-apoptotic and anti-proliferative functions, together with the androgenic regulation of RGN expression, first pointed out the role of this protein in spermatogenesis (Figure 2). The anti-oxidative properties of RGN, and its capacity suppressing oxidative damage, also have been described and may account for the maintenance of viability and fertility potential of spermatozoa from spermatogenesis until fertilization of ovum, as well as, in sperm cryopreservation (Figure 2).

Another aspect of RGN' biological actions is related to the control of Ca^{2+} homeostasis, an ion that has a crucial role in sperm function. RGN is reported to reduce intracellular Ca^{2+} concentrations in somatic cells by activation of Ca^{2+}-ATPases in mitochondrial and endoplasmic reticular membranes [128, 129] and Ca^{2+}/Mg^{2+}-ATPase in the plasma membrane [130]. The presence of RGN in spermatozoa cytoplasm and nucleus, as well as, in the acrosome along with its membrane association suggest it might be involved in the efflux of Ca^{2+} thereby reducing intracellular Ca^{2+} levels. Therefore, the anti-capacitatory function indicated for RGN may be due to Ca^{2+}- efflux, as many studies have shown that elevation of sperm intracellular Ca^{2+} (Ca^{2+} influx) is required for hyperactivation, capacitation, and acrosome reaction [131-135].

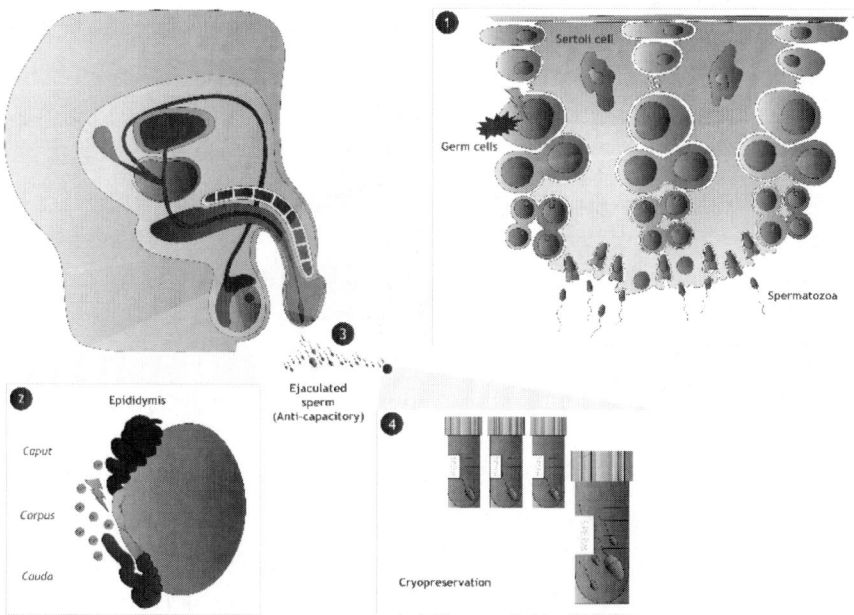

Figure 2. The broad range of RGN actions in the male reproductive tract from spermatogenesis to the fertility potential of spermatozoa. (1) RGN is expressed in testicular cells (both in germ and somatic Sertoli cells) exerting a protective function over apoptosis and oxidative damage of the germline; it also contributes to the high plasticity of Sertoli cells metabolism ensuring the supply of lactate for germ cells. As an epididymal protein (2), RGN is differentially expressed in the *caput*, *corpus* and *cauda* regions, which is related to the protection against oxidative stress and maintenance of Ca^{2+}-levels in the epididymal lumen contributing for sperm maturation. RGN also is a protein detected in the seminiferous tubules, epididymis, and seminal vesicles fluids, though it is removed from the seminal plasma after ejaculation being maintained in the acrosomal region of ejaculated sperm. Moreover, RGN seems to be an anti-capacitory agent (3) having a beneficial role in sperm cryopreservation (4), preventing premature capacitation/cryo-capacitation, Ca^{2+}-related stress, and oxidative damage.

Similar to somatic cells, Ca^{2+} entry from the extracellular space as well as its release from intracellular stores are responsible for the increase of intracellular Ca^{2+} concentrations in spermatozoa [136-140]. The internal storage of Ca^{2+} in sperm is low as mature cells do not contain endoplasmic reticulum, the major Ca^{2+} storing organelle, and potential areas for Ca^{2+} storing include the acrosome, a redundant nuclear envelope and the mitochondria packed in the midpiece [141]. Thus, the primary source of Ca^{2+} for spermatozoa is the external environment [135]. Moreover, low basal levels of intracellular Ca^{2+} are maintained by Ca^{2+} absorption by mitochondria and active Ca^{2+} extrusion by the Ca^{2+}-pump at plasma membrane [142].

In addition, the factors that regulate the function of numerous Ca^{2+}-permeable channels are known to affect male fertility [133, 143]. Disrupted Ca^{2+}-channel or –pump activity can be occurring in sperm cryopreservation, given that thawed bull and human sperm have shown to contain increased intracellular Ca^{2+} levels compared to those before cryopreservation [144, 145]. These findings reflect impaired membrane-selective permeability and/or an inability to maintain physiological concentrations of Ca^{2+} [144, 145]. Moreover, this disruption has consequences, as elevated Ca^{2+} can lead to premature acrosome reaction as well as poor fertility outcomes for the post-thaw sperm. Conversely, the proper regulation of sperm Ca^{2+} channel function can reduce the rate of premature acrosome reaction [146]. Therefore, it is likely to assume that RGN by regulating Ca^{2+} handling proteins can reduce the intracellular Ca^{2+} and minimize Ca^{2+} stress thereby protecting the spermatozoa from premature capacitation in reproductive ducts as well as from cryo-capacitation during cryopreservation.

Another possibility in preventing premature capacitation/cryo-capacitation may be due to the Ca^{2+}-binding ability of RGN. Extracellular RGN can bind Ca^{2+} ions and thus reduce intracellular Ca^{2+} by limiting the availability of extracellular Ca^{2+} that would otherwise enter the sperm through the compromised membrane pumps or cation channels. In turn, the degradation of RGN in ejaculated seminal plasma may release the bound Ca^{2+} that can be further used for capacitation. Altogether, it can be concluded that RGN has very important roles in male reproduction from spermatogenesis to fertilization. Further research would help to clarify the mechanisms underlying RGN actions in sperm physiology.

Although RGN has been assigned to many physiological functions of utmost importance in germ cell development and sperm function, no studies have been reported yet about its properties in the fertilizing ability and cryopreservation of human semen. Understanding of the mechanisms underlying the anti-capacitatory/cryoprotective effects of RGN together with the disclosure of other interacting proteins will help to define the beneficial use of RGN in assisted reproductive technologies.

REFERENCES

[1] Misawa H, Yamaguchi M (2000) The gene of Ca^{2+}-binding protein regucalcin is highly conserved in vertebrate species. *Int J Mol Med* 6: 191-196.

[2] Shimokawa N, Isogai M, Yamaguchi M (1995) Specific species and tissue differences for the gene expression of calcium-binding protein regucalcin. *Mol Cell Biochem* 143: 67-71.

[3] Goto SG (2000) Expression of Drosophila homologue of senescence marker protein-30 during cold acclimation. *J Insect Physiol* 46:1111-1120.

[4] Fujita T, Shirasawa T, Maruyama N (1996) Isolation and characterization of genomic and cDNA clones encoding mouse senescence marker protein-30 (SMP30). *Biochim Biophys Acta (BBA)-Gene Structure and Expression* 1308:49-57.

[5] Fujita T, Shirasawa T, Uchida K, Maruyama N (1992) Isolation of cDNA clone encoding rat senescence marker protein-30 (SMP30) and its tissue distribution. *Biochim Biophys Acta (BBA)-Gene Structure and Expression* 1132:297-305.

[6] Maruyama N, Ishigami A, Kondo Y (2010) Pathophysiological significance of senescence marker protein-30. *Geriatr Gerontol Int* 10 Suppl 1:S88-98.

[7] Nikapitiya C, De Zoysa M, Kang HS, Oh C, Whang I, Lee J (2008) Molecular characterization and expression analysis of regucalcin in disk abalone (Haliotis discus discus): intramuscular calcium

administration stimulates the regucalcin mRNA expression. *Comp Biochem Physiol B Biochem Mol Biol* 150:117-124.

[8] Wu YD, Jiang L, Zhou Z, Zheng MH, Zhang J, Liang Y (2008) CYP1A/regucalcin gene expression and edema formation in zebrafish embryos exposed to 2,3,7,8-Tetrachlorodibenzo-p-dioxin. *Bull Environ Contam Toxicol* 80:482-486.

[9] Nakajima Y, Natori S (2000) Identification and characterization of an anterior fat body protein in an insect. *J Biochem* 127: 901-908.

[10] Gomi K, Hirokawa K, Kajiyama N (2002) Molecular cloning and expression of the cDNAs encoding luciferin-regenerating enzyme from Luciola cruciata and Luciola lateralis. *Gene* 294: 157-166.

[11] Yamaguchi M (2000) Role of regucalcin in calcium signaling. *Life Sci* 66: 1769-1780.

[12] Yamaguchi M (2005) Role of regucalcin in maintaining cell homeostasis and function (review). *Int J Mol Med* 15:371-390.

[13] Shimokawa N, Yamaguchi M (1993) Molecular cloning and sequencing of the cDNA coding for a calcium-binding protein regucalcin from rat liver. *FEBS letters* 327:251-255.

[14] Chakraborti S, Bahnson BJ (2010) Crystal structure of human senescence marker protein 30: insights linking structural, enzymatic, and physiological functions. *Biochemistry* 49:3436-3444.

[15] Fujita T, Shirasawa T, Uchida K, Maruyama N (1996) Gene regulation of senescence marker protein-30 (SMP30): coordinated up-regulation with tissue maturation and gradual down-regulation with aging. *Mech Ageing Dev* 87:219-229.

[16] Marques R, Maia CJ, Vaz C, Correia S, Socorro S (2014) The diverse roles of calcium-binding protein regucalcin in cell biology: from tissue expression and signalling to disease. *Cell Mol Life Sci* 71:93-111.

[17] Maia CJ, Santos CR, Schmitt F, Socorro S (2008) Regucalcin is expressed in rat mammary gland and prostate and down-regulated by 17β-estradiol. *Mol Cell Biochem* 311:81-86.

[18] Laurentino SS, Correia S, Cavaco JE, Oliveira PF, Rato L, Sousa M, Barros A, Socorro S (2011) Regucalcin is broadly expressed in male

reproductive tissues and is a new androgen-target gene in mammalian testis. *Reproduction* 142:447-456.

[19] Pillai H, Shende AM, Parmar MS, Thomas J, Kartha HS, Taru Sharma G, Ghosh SK, Bhure SK (2017) Detection and localization of regucalcin in spermatozoa of water buffalo (Bubalus bubalis): A calcium-regulating multifunctional protein. *Reprod Domest Anim* 52:865-872.

[20] Pillai H, Shende AM, Parmar MS, A A, L S, Kumaresan A, G TS, Bhure SK (2017) Regucalcin is widely distributed in the male reproductive tract and exerts a suppressive effect on in vitro sperm capacitation in the water buffalo (Bubalus bubalis). *Mol Reprod Dev* 84:212-221.

[21] Correia S, Oliveira P, Guerreiro P, Lopes G, Alves M, Canário A, Cavaco J, Socorro S (2013) Sperm parameters and epididymis function in transgenic rats overexpressing the Ca^{2+}-binding protein regucalcin: a hidden role for Ca^{2+} in sperm maturation? *Mol Hum Reprod* 19:581-589.

[22] Vaz CV, Correia S, Cardoso HJ, Figueira MI, Marques R, Maia CJ, Socorro S (2016) The Emerging Role of Regucalcin as a Tumor Suppressor: Facts and Views. *Curr Mol Med* 16:607-619.

[23] Son TG, Park HR, Kim SJ, Kim K, Kim MS, Ishigami A, Handa S, Maruyama N, Chung HY, Lee J (2009) Senescence marker protein 30 is up-regulated in kainate-induced hippocampal damage through ERK-mediated astrocytosis. *J Neurosci Res* 87:2890-2897.

[24] Doran P, Dowling P, Donoghue P, Buffini M, Ohlendieck K (2006) Reduced expression of regucalcin in young and aged mdx diaphragm indicates abnormal cytosolic calcium handling in dystrophin-deficient muscle. *Biochim Biophys Acta* 1764:773-785.

[25] Lv S, Wang J-h, Liu F, Gao Y, Fei R, Du S-c, Wei L (2008) Senescence marker protein 30 in acute liver failure: validation of a mass spectrometry proteomics assay. *BMC Gastroenterol* 8:17.

[26] Jung KJ, Maruyama N, Ishigami A, Yu BP, Chung HY (2006) The redox-sensitive DNA binding sites responsible for age-related

downregulation of SMP30 by ERK pathway and reversal by calorie restriction. *Antioxid Redox Signal* 8:671-680.

[27] van Dijk KD, Berendse HW, Drukarch B, Fratantoni SA, Pham TV, Piersma SR, Huisman E, Breve JJ, Groenewegen HJ, Jimenez CR, van de Berg WD (2012) The proteome of the locus ceruleus in Parkinson's disease: relevance to pathogenesis. *Brain Pathol* 22:485-498.

[28] Jeong DH, Goo MJ, Hong IH, Yang HJ, Ki MR, Do SH, Ha JH, Lee SS, Park JK, Jeong KS (2008) Inhibition of radiation-induced apoptosis via overexpression of SMP30 in Smad3-knockout mice liver. *J Radiat Res* 49: 653-660.

[29] Fukaya Y, Yamaguchi M (2004) Regucalcin increases superoxide dismutase activity in rat liver cytosol. *Biol Pharm Bull* (Tokyo) 27:1444-1446.

[30] Vaz CV, Marques R, Maia CJ, Socorro S (2015) Aging-associated changes in oxidative stress, cell proliferation, and apoptosis are prevented in the prostate of transgenic rats overexpressing regucalcin. *Transl Res* 166: 693-705.

[31] Yamaguchi M (2013) The anti-apoptotic effect of regucalcin is mediated through multisignaling pathways. *Apoptosis* 18:1145-1153.

[32] Vaz CV, Maia CJ, Marques R, Gomes IM, Correia S, Alves MG, Cavaco JE, Oliveira PF, Socorro S (2014) Regucalcin is an androgen-target gene in the rat prostate modulating cell-cycle and apoptotic pathways. *Prostate* 74:1189-1198.

[33] Marques R, Vaz CV, Maia CJ, Gomes M, Gama A, Alves G, Santos CR, Schmitt F, Socorro S (2015) Histopathological and in vivo evidence of regucalcin as a protective molecule in mammary gland carcinogenesis. *Exp Cell Res* 330:325-335.

[34] Silva AM, Correia S, Socorro S, Maia CJ (2016) Endogenous Factors in the Recovery of Reproductive Function after Testicular Injury and Cancer. *Curr Mol Med* 16:631-649.

[35] Yamaguchi M (2015) Involvement of regucalcin as a suppressor protein in human carcinogenesis: insight into the gene therapy. *J Cancer Res Clin Oncol* 141: 1333-1341.

[36] Yamaguchi M (2010) Regucalcin and metabolic disorders: osteoporosis and hyperlipidemia are induced in regucalcin transgenic rats. *Mol Cell Biochem* 341:119-133.

[37] Vaz CV, Marques R, Cardoso HJ, Maia CJ, Socorro S (2016) Suppressed glycolytic metabolism in the prostate of transgenic rats overexpressing calcium-binding protein regucalcin underpins reduced cell proliferation. *Transgenic Res* 25:139-148.

[38] Laurentino SS, Correia S, Cavaco JE, Oliveira PF, de Sousa M, Barros A, Socorro S (2012) Regucalcin, a calcium-binding protein with a role in male reproduction? *Mol Hum Reprod* 18:161-170.

[39] Yamaguchi M (1988) Physicochemical properties of calcium-binding protein isolated from rat liver cytosol: Ca^{2+}-induced conformational changes. *Chem Pharm Bull (Tokyo)* 36: 286-290.

[40] Harikrishna P, Thomas J, Shende AM, Bhure SK (2017) Calcium Binding Ability of Recombinant Buffalo Regucalcin: A Study Using Circular Dichroism Spectroscopy. *Protein J* 36:108-111.

[41] Kondo Y, Inai Y, Sato Y, Handa S, Kubo S, Shimokado K, Goto S, Nishikimi M, Maruyama N, Ishigami A (2006) Senescence marker protein 30 functions as gluconolactonase in L-ascorbic acid biosynthesis, and its knockout mice are prone to scurvy. *Proc Natl Acad Sci U S A* 103:5723-5728.

[42] Ishigami A, Handa S, Maruyama N, Supakar PC (2003) Nuclear localization of senescence marker protein-30, SMP30, in cultured mouse hepatocytes and its similarity to RNA polymerase. *Biosci Biotechnol Biochem* 67:158-160.

[43] Tsurusaki Y, Misawa H, Yamaguchi M (2000) Translocation of regucalcin to rat liver nucleus: involvement of nuclear protein kinase and protein phosphatase regulation. *Int J Mol Med* 6:655-715.

[44] Morooka Y, Yamaguchi M (2002) Endogenous regucalcin suppresses the enhancement of protein phosphatase activity in the

cytosol and nucleus of kidney cortex in calcium-administered rats. *J Cell Biochem* 85:553-560.

[45] Tobisawa M, Yamaguchi M (2003) Role of endogenous regucalcin in brain function: suppression of cytosolic nitric oxide synthase and nuclear protein tyrosine phosphatase activities in brain tissue of transgenic rats. *Int J Mol Med* 12:581-585.

[46] Ichikawa E, Tsurusaki Y, Yamaguchi M (2004) Suppressive effect of regucalcin on protein phosphatase activity in the heart cytosol of normal and regucalcin transgenic rats. *Int J Mol Med* 13:289-293.

[47] Arun P, Aleti V, Parikh K, Manne V, Chilukuri N (2011) Senescence marker protein 30 (SMP30) expression in eukaryotic cells: existence of multiple species and membrane localization. *PLoS One* 6: e16545.

[48] Omura M, Yamaguchi M (1999) Regulation of protein phosphatase activity by regucalcin localization in rat liver nuclei. *J Cell Biochem* 75: 437-445.

[49] Nakagawa T, Yamaguchi M (2006) Overexpression of regucalcin enhances its nuclear localization and suppresses L-type Ca^{2+} channel and calcium-sensing receptor mRNA expressions in cloned normal rat kidney proximal tubular epithelial NRK52E cells. *J Cell Biochem* 99:1064-1077.

[50] Nakagawa T, Yamaguchi M (2008) Nuclear localization of regucalcin is enhanced in culture with protein kinase C activation in cloned normal rat kidney proximal tubular epithelial NRK52E cells. *Int J Mol Med* 21: 605-610.

[51] Maia C, Santos C, Schmitt F, Socorro S (2009) Regucalcin is under-expressed in human breast and prostate cancers: Effect of sex steroid hormones. *J Cell Biochem* 107:667-676.

[52] Vaz CV, Rodrigues DB, Socorro S, Maia CJ (2015) Effect of extracellular calcium on regucalcin expression and cell viability in neoplastic and non-neoplastic human prostate cells. *Biochim Biophys Acta* 1853(10 Pt A): 2621-2628.

[53] Morooka Y, Yamaguchi M (2001) Inhibitory effect of regucalcin on protein phosphatase activity in the nuclei of rat kidney cortex. *J Cell Biochem* 83:111-120.

[54] Shimokawa N, Matsuda Y, Yamaguchi M (1995) Genomic cloning and chromosomal assignment of rat regucalcin gene. *Mol Cell Biochem* 151: 157-163.

[55] Fujita T, Mandel JL, Shirasawa T, Hino O, Shirai T, Maruyama N (1995) Isolation of cDNA clone encoding human homologue of senescence marker protein-30 (SMP30) and its location on the X chromosome. *Biochim Biophys Acta* 1263:249-252.

[56] Shamsi MB, Kumar K, Dada R (2011) Genetic and epigenetic factors: Role in male infertility. *Indian J Urol* 27:110-120.

[57] Brown CJ, Goss SJ, Lubahn DB, Joseph DR, Wilson EM, French FS, Willard HF (1989) Androgen receptor locus on the human X chromosome: regional localization to Xq11-12 and description of a DNA polymorphism. *Am J Hum Genet* 44:264-269.

[58] Ferlin A, Vinanzi C, Garolla A, Selice R, Zuccarello D, Cazzadore C, Foresta C (2006) Male infertility and androgen receptor gene mutations: clinical features and identification of seven novel mutations. *Clin Endocrinol (Oxf)* 65: 606-610.

[59] O'Hara L, Smith LB (2017) The Genetics of Androgen Receptor Signalling in Male Fertility. In: *Genetics of Human Infertility*. P.H. Vogt, Editor: Basel, Karger. p. 86-100.

[60] Stouffs K, Tournaye H, Liebaers I, Lissens W (2009) Male infertility and the involvement of the X chromosome. *Hum Reprod Update* 15:623-637.

[61] Zheng K, Yang F, Wang PJ (2010) Regulation of Male Fertility by X-Linked Genes. *J Androl* 31:79-85.

[62] Kosugi S, Hasebe M, Tomita M, Yanagawa H (2009) Systematic identification of cell cycle-dependent yeast nucleocytoplasmic shuttling proteins by prediction of composite motifs. *Proc Natl Acad Sci U S A* 106: 10171-10176.

[63] Sievers F, Wilm A, Dineen D, Gibson TJ, Karplus K, Li W, Lopez R, McWilliam H, Remmert M, Soding J, Thompson JD, Higgins DG

(2011) Fast, scalable generation of high-quality protein multiple sequence alignments using Clustal Omega. *Mol Syst Biol* 7: 539.

[64] Holdcraft RW, Braun RE (2004) Hormonal regulation of spermatogenesis. *Int J Androl* 27:335-342.

[65] Walker WH (2009) Molecular mechanisms of testosterone action in spermatogenesis. *Steroids* 74:602-607.

[66] Walker WH, Cheng J (2005) FSH and testosterone signaling in Sertoli cells. *Reproduction* 130:15-28.

[67] Zhou X (2010) Roles of Androgen Receptor in Male and Female Reproduction: Lessons From Global and Cell-Specific Androgen Receptor Knockout (ARKO) Mice. *J Androl* 31:235-243.

[68] Verhoeven G, Willems A, Denolet E, Swinnen JV, De Gendt K (2010) Androgens and spermatogenesis: lessons from transgenic mouse models. *Philos Trans R Soc Lond B Biol Sci* 365:1537-1556.

[69] Oliveira PF, Alves MG, Martins AD, Correia S, Bernardino RL, Silva J, Barros A, Sousa M, Cavaco JE, Socorro S (2014) Expression pattern of G protein-coupled receptor 30 in human seminiferous tubular cells. *Gen Comp Endocrinol* 201:16-20.

[70] Cavaco JEB, Laurentino SS, Barros A, Sousa M, Socorro S (2009) Estrogen receptors α and β in human testis: both isoforms are expressed. *Syst Biol Reprod Med* 55:137-144.

[71] Correia S, Cardoso HJ, Cavaco JE, Socorro S (2015) Oestrogens as apoptosis regulators in mammalian testis: angels or devils? *Expert Rev Mol Med* 17:e2.

[72] Correia S, Alves MR, Cavaco JE, Oliveira PF, Socorro S (2014) Estrogenic regulation of testicular expression of stem cell factor and c-kit: implications in germ cell survival and male fertility. *Fertil Steril* 102: 299-306.

[73] Hewitt SC, Korach KS (2018) Estrogen Receptors: New Directions in the New Millennium. *Endocr Rev* 39:664-675.

[74] Pinto P, Estêvão M, Socorro S (2014) Androgen receptors in non-mammalian vertebrates: Structure and function. In: *Androgen Receptor: Structural Biology, Genetics and Molecular Defects.* Nova Science Publishers, Inc: New York, USA. p. 1-26.

[75] Figueira M, Cardoso H, Socorro S (2018) The Role of GPER Signaling in Carcinogenesis: a Focus on Prostate Cancer. In: *Recent Trends in Cancer Biology: Spotlight on Signaling Cascades and microRNAs*. Springer: New York, USA. p. 59-117.

[76] Socorro S, Power DM, Olsson PE, Canario AV (2000) Two estrogen receptors expressed in the teleost fish, Sparus aurata: cDNA cloning, characterization and tissue distribution. *J Endocrinol* 166:293-306.

[77] Laurentino S, Gonçalves J, Cavaco JE, Oliveira PF, Alves MG, de Sousa M, Barros A, Socorro S (2011) Apoptosis-inhibitor Aven is downregulated in defective spermatogenesis and a novel estrogen target gene in mammalian testis. *Fertil Steril* 96:745-750.

[78] Correia S, Vaz CV, Silva A, Cavaco JE, Socorro S (2017) Regucalcin counteracts tert-butyl hydroperoxide and cadmium-induced oxidative stress in rat testis. *J Appl Toxicol* 37:159-166.

[79] Yamaguchi M, Oishi K (1995) 17 beta-Estradiol stimulates the expression of hepatic calcium-binding protein regucalcin mRNA in rats. *Mol Cell Biochem* 143:137-141.

[80] Kurota H, Yamaguchi M (1996) Steroid hormonal regulation of calcium-binding protein regucalcin mRNA expression in the kidney cortex of rats. *Mol Cell Biochem* 155:105-111.

[81] Cucuzza LS, Divari S, Mulasso C, Biolatti B, Cannizzo FT (2014) Regucalcin expression in bovine tissues and its regulation by sex steroid hormones in accessory sex glands. *PLoS One* 9:e113950.

[82] Correia S (2014) *Estrogens and regucalcin in testicular apoptosis and sperm function: "a matter of life and death."* University of Beira Interior.

[83] Lin PH, Jian CY, Chou JC, Chen CW, Chen CC, Soong C, Hu S, Lieu FK, Wang PS, Wang SW (2016) Induction of renal senescence marker protein-30 (SMP30) expression by testosterone and its contribution to urinary calcium absorption in male rats. *Sci Rep* 6:32085.

[84] Gorczynska E, Handelsman DJ (1995) Androgens rapidly increase the cytosolic calcium concentration in Sertoli cells. *Endocrinology* 136:2052-2059.

[85] Audy MC, Vacher P, Duly B (1996) 17 beta-estradiol stimulates a rapid Ca^{2+} influx in LNCaP human prostate cancer cells. *Eur J Endocrinol* 135: 367-373.

[86] Picotto G, Vazquez G, Boland R (1999) 17beta-oestradiol increases intracellular Ca^{2+} concentration in rat enterocytes. Potential role of phospholipase C-dependent store-operated Ca^{2+} influx. *Biochem J* 339 (Pt 1): 71-77.

[87] Azenabor AA, Hoffman-Goetz L (2001) 17 beta-estradiol increases $Ca(^{2+})$ influx and down regulates interleukin-2 receptor in mouse thymocytes. *Biochem Biophys Res Commun* 281:277-281.

[88] Sun YH, Gao X, Tang YJ, Xu CL, Wang LH (2006) Androgens induce increases in intracellular calcium via a G protein-coupled receptor in LNCaP prostate cancer cells. *J Androl* 27:671-678.

[89] Fayad T, Lévesque V, Sirois J, Silversides DW, Lussier JG (2004) Gene expression profiling of differentially expressed genes in granulosa cells of bovine dominant follicles using suppression subtractive hybridization. *Biol Reprod* 70:523-533.

[90] Mori T, Ishigami A, Seyama K, Onai R, Kubo S, Shimizu K, Maruyama N, Fukuchi Y (2004) Senescence marker protein-30 knockout mouse as a novel murine model of senile lung. *Pathol Int* 54:167-173.

[91] Print CG, Loveland KL (2000) Germ cell suicide: new insights into apoptosis during spermatogenesis. *Bioessays* 22:423-430.

[92] Lin WW, Lamb DJ, Wheeler TM, Abrams J, Lipshultz LI, Kim ED (1997) Apoptotic frequency is increased in spermatogenic maturation arrest and hypospermatogenic states. *J Urol* 158:1791-1793.

[93] Takagi S, Itoh N, Kimura M, Sasao T, Tsukamoto T (2001) Spermatogonial proliferation and apoptosis in hypospermatogenesis associated with nonobstructive azoospermia. *Fertil Steril* 76:901-907.

[94] Bozec A, Amara S, Guarmit B, Selva J, Albert M, Rollet J, El Sirkasi M, Vialard F, Bailly M, Benahmed M (2008) Status of the

executioner step of apoptosis in human with normal spermatogenesis and azoospermia. *Fertil Steril* 90:1723-1731.

[95] Correia S, Alves MG, Oliveira PF, Alves MR, van Pelt AM, Cavaco JE, Socorro S (2014) Transgenic overexpression of regucalcin leads to suppression of thapsigargin- and actinomycin D-induced apoptosis in the testis by modulation of apoptotic pathways. *Andrology* 2:290-298.

[96] Silva AM, Correia S, Casalta-Lopes JE, Mamede AC, Cavaco JE, Botelho MF, Socorro S, Maia CJ (2016) The protective effect of regucalcin against radiation-induced damage in testicular cells. *Life Sci* 164:31-41.

[97] Yamaguchi M, Morooka Y, Misawa H, Tsurusaki Y, Nakajima R (2002) Role of endogenous regucalcin in transgenic rats: Suppression of kidney cortex cytosolic protein phosphatase activity and enhancement of heart muscle microsomal Ca^{2+}-ATPase activity. *J Cell Biochem* 86:520-529.

[98] Gervasi MG, Visconti PE (2017) Molecular changes and signaling events occurring in spermatozoa during epididymal maturation. *Andrology* 5: 204-218.

[99] Dacheux J-L, Belleannée C, Jones R, Labas V, Belghazi M, Guyonnet B, Druart X, Gatti JL, Dacheux F (2009) Mammalian epididymal proteome. *Mol Cell Endocrinol* 306:45-50.

[100] Robaire B, Hinton BT, Orgebin-Crist M-C (2006) The epididymis. In: *Knobil and Neil´s Physiology of Reproduction*. Elsevier: San Diego, CA. p. 1071-1148.

[101] Aitken RJ, Curry BJ (2011) Redox regulation of human sperm function: from the physiological control of sperm capacitation to the etiology of infertility and DNA damage in the germ line. *Antioxid Redo Signal* 14: 367-381.

[102] Handa S, Maruyama N, Ishigami A (2009) Over-expression of Senescence Marker Protein-30 decreases reactive oxygen species in human hepatic carcinoma Hep G2 cells. *Biol Pharm Bull (Tokyo)* 32:1645-1648.

[103] Ichikawa E, Yamaguchi M (2004) Regucalcin increases superoxide dismutase activity in the heart cytosol of normal and regucalcin transgenic rats. *Int J Mol Med* 14:691-696.

[104] Son TG, Zou Y, Jung KJ, Yu BP, Ishigami A, Maruyama N, Lee J (2006) SMP30 deficiency causes increased oxidative stress in brain. *Mech Ageing Dev* 127:451-457.

[105] Sato T, Seyama K, Sato Y, Mori H, Souma S, Akiyoshi T, Kodama Y, Mori T, Goto S, Takahashi K, Fukuchi Y, Maruyama N, Ishigami A (2006) Senescence marker protein-30 protects mice lungs from oxidative stress, aging, and smoking. *Am J Respir Crit Care Med* 174:530-537.

[106] Alves MG, Rato L, Carvalho RA, Moreira PI, Socorro S, Oliveira PF (2013) Hormonal control of Sertoli cell metabolism regulates spermatogenesis. *Cell Mol Life Sci* 70:777-793.

[107] Mateus I, Feijo M, Espinola LM, Vaz CV, Correia S, Socorro S (2018) Glucose and glutamine handling in the Sertoli cells of transgenic rats overexpressing regucalcin: plasticity towards lactate production. *Sci Rep* 8:10321.

[108] Maneesh M, Jayalekshmi H (2006) Role of reactive oxygen species and antioxidants on pathophysiology of male reproduction. *Indian J Clin Biochem* 21:80-89.

[109] Sato Y, Kajiyama S, Amano A, Kondo Y, Sasaki T, Handa S, Takahashi R, Fukui M, Hasegawa G, Nakamura N, Fujinawa H, Mori T, Ohta M, Obayashi H, Maruyama N, Ishigami A (2008) Hydrogen-rich pure water prevents superoxide formation in brain slices of vitamin C-depleted SMP30/GNL knockout mice. *Biochem Biophys Res Commun* 375:346-350.

[110] Kondo Y, Sasaki T, Sato Y, Amano A, Aizawa S, Iwama M, Handa S, Shimada N, Fukuda M, Akita M (2008) Vitamin C depletion increases superoxide generation in brains of SMP30/GNL knockout mice. *Biochem Biophys Res Commun* 377:291-296.

[111] Kim HS, Son TG, Park HR, Lee Y, Jung Y, Ishigami A, Lee J (2013) Senescence marker protein 30 deficiency increases

Parkinson's pathology by impairing astrocyte activation. *Neurobiol Aging* 34:1177-1183.

[112] Son TG, Kim SJ, Kim K, Kim MS, Chung HY, Lee J (2008) Cytoprotective roles of senescence marker protein 30 against intracellular calcium elevation and oxidative stress. *Arch Pharm Res* 31:872-877.

[113] Linster CL, Van Schaftingen E (2007) Vitamin C. Biosynthesis, recycling and degradation in mammals. *Febs J* 274:1-22.

[114] Paudel KP, Kumar S, Meur SK, Kumaresan A (2010) Ascorbic acid, catalase and chlorpromazine reduce cryopreservation-induced damages to crossbred bull spermatozoa. *Reprod Domest Anim* 45:256-262.

[115] Puga Molina LC, Luque GM, Balestrini PA, Marin-Briggiler CI, Romarowski A, Buffone MG (2018) Molecular Basis of Human Sperm Capacitation. *Front Cell Dev Biol* 6:72.

[116] Desnoyers L, Manjunath P (1992) Major proteins of bovine seminal plasma exhibit novel interactions with phospholipid. *J Biol Chem* 267: 10149-10155.

[117] de Lamirande E, Yoshida K, Yoshiike TM, Iwamoto T, Gagnon C (2001) Semenogelin, the main protein of semen coagulum, inhibits human sperm capacitation by interfering with the superoxide anion generated during this process. *J Androl* 22:672-679.

[118] Cross NL (1996) Human seminal plasma prevents sperm from becoming acrosomally responsive to the agonist, progesterone: cholesterol is the major inhibitor. *Biol Reprod* 54:138-145.

[119] Andrews JC, Nolan JP, Hammerstedt RH, Bavister BD (1994) Role of zinc during hamster sperm capacitation. *Biol Reprod* 51:1238-1247.

[120] Tomes CN, Carballada R, Moses DF, Katz DF, Saling PM (1998) Treatment of human spermatozoa with seminal plasma inhibits protein tyrosine phosphorylation. *Mol Hum Reprod* 4:17-25.

[121] Hirohashi N, Yanagimachi R (2018) Sperm acrosome reaction: its site and role in fertilization. *Biol Reprod* 99:127-133.

[122] Bailey JL, Bilodeau JF, Cormier N (2000) Semen cryopreservation in domestic animals: a damaging and capacitating phenomenon. *J Androl* 21: 1-7.

[123] Hezavehei M, Sharafi M, Kouchesfahani HM, Henkel R, Agarwal A, Esmaeili V, Shahverdi A (2018) Sperm cryopreservation: A review on current molecular cryobiology and advanced approaches. *Reprod Biomed Online* 37:327-339.

[124] Said TM, Grunewald S, Paasch U, Rasch M, Agarwal A, Glander HJ (2005) Effects of magnetic-activated cell sorting on sperm motility and cryosurvival rates. *Fertil Steril* 83:1442-1446.

[125] Chatterjee S, de Lamirande E, Gagnon C (2001) Cryopreservation alters membrane sulfhydryl status of bull spermatozoa: protection by oxidized glutathione. *Mol Reprod Dev* 60:498-506.

[126] Vesala L, Salminen TS, Kankare M, Hoikkala A (2012) Photoperiodic regulation of cold tolerance and expression levels of regucalcin gene in Drosophila montana. *J Insect Physiol* 58:704-709.

[127] Pillai H, Parmar MS, Shende AM, Thomas J, Sharma HS, Sharma GT, Ghosh SK, Kumaresan A, Bhure SK (2017) Effect of supplementation of recombinant Regucalcin in extender on cryopreservation of spermatozoa of water buffalo (Bubalus bubalis). *Mol Reprod Dev* 84:1133-1139.

[128] Takahashi H, Yamaguchi M (1999) Role of regucalcin as an activator of Ca^{2+}-ATPase activity in rat liver microsomes. *J Cell Biochem* 74: 663-669.

[129] Yamaguchi M, Mori S (1989) Activation of hepatic microsomal Ca^{2+}-adenosine triphosphatase by calcium-binding protein regucalcin. *Chem Pharm Bul (Tokyo)l* 37:1031-1034.

[130] Takahashi H, Yamaguchi M (1993) Regulatory effect of regucalcin on (Ca^{2+}– Mg^{2+})-ATPase in rat liver plasma membranes: comparison with the activation by Mn^{2+} and Co^{2+}. *Mol Cell Biochem* 124:169-174.

[131] Mannowetz N, Naidoo NM, Choo SA, Smith JF, Lishko PV (2013) Slo1 is the principal potassium channel of human spermatozoa. *Elife* 2: e01009.

[132] Kwon WS, Park YJ, Mohamed el SA, Pang MG (2013) Voltage-dependent anion channels are a key factor of male fertility. *Fertil Steril* 99:354-361.

[133] Lishko PV, Kirichok Y (2010) The role of Hv1 and CatSper channels in sperm activation. *J Physiol* 588:4667-472.

[134] Suarez SS (2008) Control of hyperactivation in sperm. *Hum Reprod Update* 14:647-657.

[135] Foresta C, Rossato M (1997) Calcium influx pathways in human spermatozoa. *Mol Hum Reprod* 3:1-4.

[136] Benoff S, Chu CC, Marmar JL, Sokol RZ, Goodwin LO, Hurley IR (2007) Voltage-dependent calcium channels in mammalian spermatozoa revisited. *Front Biosci* 12:1420-1449.

[137] Breitbart H (2002) Intracellular calcium regulation in sperm capacitation and acrosomal reaction. *Mol Cell Endocrinol* 187:139-144.

[138] Darszon A, Lopez-Martinez P, Acevedo JJ, Hernandez-Cruz A, Trevino CL (2006) T-type Ca2+ channels in sperm function. *Cell Calcium* 40: 241-252.

[139] Florman HM, Jungnickel MK, Sutton KA (2008) Regulating the acrosome reaction. *Int J Dev Biol* 52:503-510.

[140] Publicover SJ, Barratt CL (1999) Voltage-operated Ca^{2+} channels and the acrosome reaction: which channels are present and what do they do? *Hum Reprod* 14:873-879.

[141] Costello S, Michelangeli F, Nash K, Lefievre L, Morris J, Machado-Oliveira G, Barratt C, Kirkman-Brown J, Publicover S (2009) Ca^{2+}-stores in sperm: their identities and functions. *Reproduction* 138:425-437.

[142] Wennemuth G, Babcock DF, Hille B (2003) Calcium clearance mechanisms of mouse sperm. *J Gen Physiol* 122:115-128.

[143] Ren D, Xia J (2010) Calcium signaling through CatSper channels in mammalian fertilization. *Physiology* 25:165-175.

[144] Bailey JL, Buhr MM (1993) Ca^{2+} regulation by cryopreserved bull spermatozoa in response to A23187. *Cryobiology* 30:470-481.

[145] McLaughlin EA, Ford WC (1994) Effects of cryopreservation on the intracellular calcium concentration of human spermatozoa and its response to progesterone. *Mol Reprod Dev* 37:241-246.

[146] Florman HM, Arnoult C, Kazam IG, Li C, O'Toole CM (1998) A perspective on the control of mammalian fertilization by egg-activated ion channels in sperm: a tale of two channels. *Biol Reprod* 59:12-16.

In: Regucalcin
Editor: Masayoshi Yamaguchi

ISBN: 978-1-53616-172-4
© 2019 Nova Science Publishers, Inc.

Chapter 3

REGUCALCIN: A TUMOR-ASSOCIATED ANTIGEN AND A TUMOR-SUPPRESSING MOLECULE

Su-Fang Zhou[*]
Department of Biochemistry and Molecular Biology,
Guangxi Medical University, Nanning, Guangxi, P.R. China

ABSTRACT

Regucalcin is a multifunctional molecule and various studies have suggested that it is closely related to the development of tumors. It has been found that anti-regucalcin antibodies are present in the serum of some tumor patients such as liver cancer sufferers. The anti-regucalcin antibody positive rate is higher in AFP-negative liver cancer serum than that in AFP-positive ones. The frequency of serum antibody against regucalcin with well histopathologically differentiated samples is significantly higher than that in poorly differentiated ones. This indicates that regucalcin is a tumor-associated antigen and it can be used as a serum marker for tumor diagnosis. Regucalcin is expressed in normal tissues in the body, especially in liver and kidney. It is, however,

---

[*] Corresponding Author's E-mail: zsf200000@163.com.

generally showing low expression in tumor tissues, which is closely related to the shorter survival rate of patients. The low expression level of regucalcin in liver cancer cells and liver cancer tissues is inversely proportional to the frequency of methylation of its gene promoter. *In vitro* experiments have shown that silencing regucalcin in tumor cells will promote tumor cell invasion and migration. Regucalcin inhibits the proliferation of tumor cells by inhibiting the expression of multiple protein kinases, increasing the expression of tumor suppressor genes, reducing oncogene expression and regulating apoptosis-related proteins with independent of cell death. Those evidences suggest that regucalcin is a tumor-suppressing molecule and its level of expression can be used as a prognostic indicator for patients.

Keywords: regucalcin, tumor-associated antigen, anti-regucalcin antibody, tumor-suppressor

INTRODUCTION

Cancer patients can generate humoral- and cell-mediated immune response to abnormal proteins expressed in tumors and these proteins are called tumor-associated antigens (TAAs). Many of these TAAs have been shown to have potentiality for application in therapy, diagnosis and monitor of tumor progress. Tumor-associated autoantibodies (AAbs) are produced as an immune response to the TAAs. Those AAbs represent one novel pathway of early detection markers for cancer.

Regucalcin (RGN) is a tumor-associated antigen, which was first identified from Guangxi human hepatocellular carcinoma (HCC) cDNA expression library by a serological analysis of recombinant cDNA expression library (SEREX) approach. The results of this study indicate that there are antibodies against regucalcin in the serum of some cancer patients, mainly found in the serum of patients with liver cancer, especially those with negative alpha-fetoprotein (AFP). The frequency of serum antibody against regucalcin is significantly higher in histopathologically diagnosed as a well differentiated hepatocellular carcinoma than that in poorly differentiated ones. These findings demonstrate that the regucalcin-associated autoantibodies can be used as a marker to scale up frequency analysis of serum antibody against cancer and to improve the understanding of relevancy of autoantibody with tumor. The mechanism of the antibody production to regucalcin still requires in-depth research.

In recent years, more and more researchers have noticed that RGN is related to cancer development. Several reports have demonstrated that the expression of regucalcin has decreased in tumor tissues and this is closely related to the tumor cell invasion and migration and the shorter survival rate of patients. Those evidences suggest that RGN is a tumor suppressing molecule and its level of expression can be used as a prognostic indicator for patients. The expression of regucalcin gene is regulated through various hormone, transcription factors and epigenetic alterations. The experiments *in vitro* have shown that regucalcin inhibits tumorigenesis through a variety of signaling pathways.

This chapter is written trying to summarize the advances in understanding the dual characteristics of regucalcin during the process of cancer development, namely as a tumor-associated antigen and as a tumor-suppressor.

REGUCALCIN: A TUMOR-ASSOCIATED ANTIGEN

Tumor-Associated Antigens and Their Autoantibodies

Tumor-associated antigens are antigenic substances aberrantly expressed, mutated or posttranslationally modified proteins or other autoantigens associated with tumors and may be enhanced by tumor-associated inflammation [1]. A link between autoimmune responses and cancer via autoantibodies was first described in the 1950s. The TAAs were first identified in melanoma patients and demonstrated by the identification of MAGEA1 specific T cell clones in the tumor infiltrating lymphocytes. Many such TAAs have now been identified. One of the most prominent being members of the cancer-testis (CT) gene family and include antigens such as MAGEA3 and NY-ESO-1. Other TAAs include differentiation antigens (e.g., Melan-A/MART-1), altered antigens (e.g., MUC-1) and over expressed antigens (e.g., Her2-Neu) or oncoviral antigens (e.g., HPV, EBV). What is significant is that it triggers an immune response in the host. Host T cells are capable of recognizing these epitopes, indicating that humoral immunity and cellular immunity can be induced spontaneously so that it is feasible to break tolerance [2].

Tumor-associated autoantibodies are produced as an immune response to the TAAs. Those AAbs represent one novel pathway of early detection markers for cancer. Over the last few decades, AAbs have become of particularly attractive as cancer biomarkers because they can be easily extracted from serum via minimally invasive blood collection. They also exhibit increased levels in very early cancer stages and are observed in patients with various carcinomas. The AAbs are particularly attractive as diagnostic biomarkers for cancer given that they may circulate at higher concentrations than their corresponding antigen and demonstrate higher stability over time. So their production may precede clinical confirmation of a tumor by several months or years. Additionally, they persist for extended periods after the corresponding antigen is no longer detectable, at lasting concentrations and with long half-lives in blood, due to limited proteolysis and clearance from the circulation. Those characteristics make sample handling less arduous [1].

The searching for AAbs has been accelerated by the advent of SEREX, cancer genome sequencing, serological proteome analysis (SERPA), 2D immunoblotting and mass spectrometry. This has led to increasing numbers of AAbs for which elevated serum levels have been found in cancer patients for a variety of cancer types [3-6], including liver, breast, lung, gastrointestinal, ovarian and prostate cancer. The exact factors that contribute to an enhancement or disturbance of immune surveillance leading to the production of autoantibodies in cancer are still illusive. The question remains as how and why cellular components may be rendered immunogenic in cancer. Figure 1 summarizes some of the major theories surrounding the production of autoantibodies in cancer. They include loss of tolerance, inflammation, changes in antigen expression as well as their altered exposure or altered presentation, reduced degradation, posttranslational modifications (PTMs) and their aberrant location or altered structure [1].

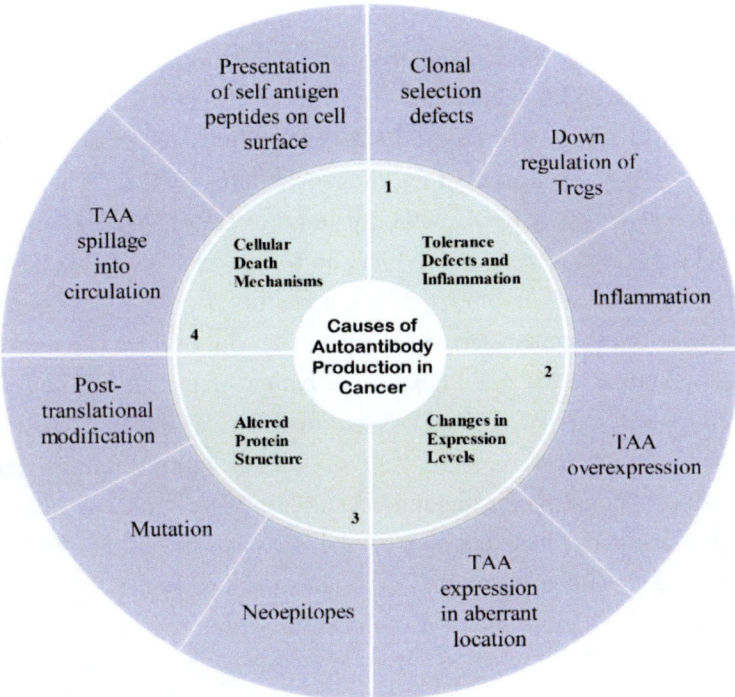

Figure 1. The proposed causes of autoantibody production in cancer [1].

Identification of Regucalcin as a Tumor-Associated Antigen by Using SEREX in Hepatocellular Carcinoma

The SEREX is a method, first developed by the group of Sahin U. in 1995, used to analyze tumor-associated antigenic structures by examining antibody responses in the tumor-bearing host. It utilizes the sera of cancer patients, which contain antibodies against a variety of tumor antigens, to screen for tumor antigens in cDNA expression libraries constructed from tumor tissue or cell lines [7], as illustrated in Figure 2. Some tumor antigens have been identified by conventional serology previously. With the introduction of SEREX, however, the amount of antibody detectable antigens has risen rapidly. The broad applicability with respect to different tumor entities and the fast access to molecular characterization are the major advantages of this method [8].

By applying the SEREX method, such as cloning, HCC-22-5 has been isolated as an HCC-associated antigen from a Guangxi HCC cDNA library screening with autogenous serum [9, 10]. In order to comprehend immune response of HCC-22-5 in distinct allogeneic sera, HCC-22-5 antigen has been tested again by SEREX method using the sera of HCC, glioblastoma, melanocytoma, renal cell carcinoma, non-hodgkin's lymphoma, ovarian cancer, hepatitis B&C, liver cirrhosis and healthy people. It has been found from the above allogeneic serum samples that 3 out of 5 HCC patients, 1 in 20 healthy people and 1 in 10 hepatitis B patients have shown high-tittered IgG antibodies against HCC-22-5 antigen. In 20 patients with other malignant tumors, however, 10 patients with hepatitis C and 4 patients with cirrhosis, no antibodies against HCC-22-5 were detected. This cloned gene of antigens has been sequenced and is shown having 860 nucleotide, in which 500 nucleotides are in open reading frame of HCC-22-5 representing COOH-terminal 165 amino acids of regucalcin.

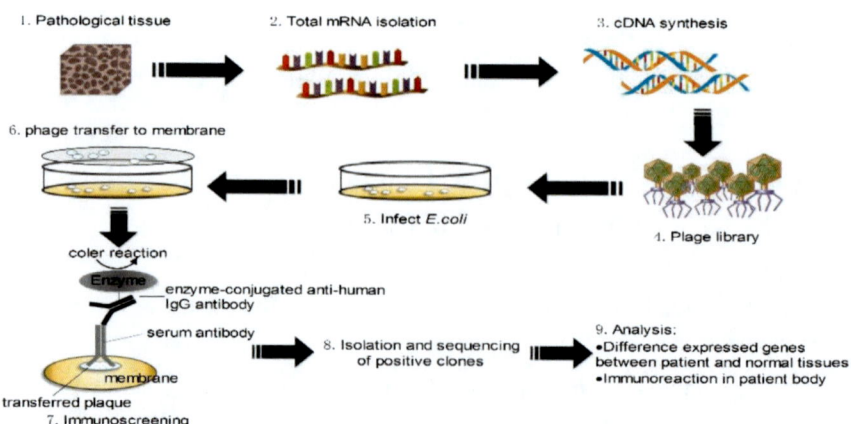

Figure 2. A diagram showing procedures involved in the SEREX technology [7].

The above results have demonstrated that regucalcin, an HCC-associated antigen, can be used as a marker to scale up frequency analysis of serum antibody against HCC. But the mechanism of how this autoantibody is relevant to HCC is still not fully understood.

Anti-Regucalcin Antibody Detected in Tumor Patients

The regucalcin gene located at chromosome X can abundantly express in the hepatocytes and is important to the functionalities of liver [11, 12]. As mentioned earlier, the use of SEREX technology can detect the presence of anti-regucalcin in the serum of patients with liver cancer. In order to get a better understanding of the significance of anti-regucalcin antibody in HCC, to identify regucalcin autoantibody as a new HCC marker and to broaden our knowledge of the relationship between regucalcin seropositivity and the immune response in tumor, it is necessary to detect anti-regucalcin antibody in blood.

At the present, serum antibody detection mainly adopts the ELISA method. Comparing to the SEREX technology, the ELISA method is simple and convenient to operate.

The method of ELISA for detecting serum anti-regucalcin antibody is briefly described as follows: first of all, regucalcin recombinant protein is expressed and purified; secondly, serum antibodies are detected by ELISA; after that, cutoff values for serum antibody titers are determined. For purifying recombinant regucalcin protein, the cDNA of regucalcin is amplified via polymerase chain reaction. The gene is inserted into a plasmid expressed as a tag and the recombinant protein is expressed in *Escherichia coli* and dissolved in phosphate-buffered saline (PBS). Regucalcin extract is then added to affinity chromatography column. The purified regucalcin recombinant protein is washed and eluted with suitable buffer. The expression and purity of the recombinant proteins are examined with 12% SDS-PAGE. The DNA sequencing analysis is used to confirm whether the correct gene have been inserted into the constructed plasmid [10, 13]. The serum samples from patients and healthy controls are analyzed via ELISA. Regucalcin signals are evaluated by calculating the difference in absorbance between wells containing regucalcin and those containing PBS. Optimized antibody titer cutoff values and a standard cutoff value corresponding to a value greater than mean + 3 standard deviations (SDs) of the healthy control cohorts are used for each of the antibodies while maintaining a specificity of >95%. The specificity of the assay is determined as the percentage of healthy controls showing negative results.

Based on current research reports, the specificity and sensitivity of anti-regucalcin antibody in the serum of patients with cancer by ELISA are shown in Table 1 and Figure 3.

Table 1. Frequency of autoantibodies to regucalcin in various diseases [15-16]

Group	n	Sensitivity	specificity
Hepatocellular carcinoma	145	32.4	95.8
nasopharyngeal carcinoma	98	13.3	95.8
Hepatitis B	60	6.6	95.8
cancer of cervix	41	7.3	95.8
esophageal cancer	26	7.7	95.8
Gastric cancer	100	7.0	98.7
Liver cirrhosis	19	5.3	95.8
intestines cancer	40	2.5	95.8
metastasis liver cancer	12	0	95.8
carcinoma of pancreas	5	0	95.8
lung cancer	30	0	95.8
breast cancer	30	0	95.8
brain tumor	34	0	95.8
malignant lymphoma	20	0	95.8
hypertension, DM etc.	20	0	95.8
normal controls	138	0.72	95.8

Experiments have shown that the positive rate of anti-regucalcin antibody in 145 HCC patients of different age is 9.1% for younger than 30 years olds, 33.3% for 30~60 years olds and 38.5% for over 60s. There is no significant correlation of age with the frequency of antibody response, although there are reports showing that regucalcin is an age-associated protein and its expression is decreased with aging in a sex-independent manner [14].

The frequency of antibody to regucalcin, moreover, is not significantly correlated with tumor size, lymph node metastasis, tumor invasion or TNM classification. Those results indicate that the regucalcin antibody occurrence is not involved in the malignant progression of HCC. But the positive rate of regucalcin antibody is 52.6% (10/19) in HCC with pathological grade I-II and 19.0% (4/21) in grade III-IV (P<0.05). On the other hand, 78.6% (11/14) HCC patients with the positive antibody have a trabecular histological pattern.

Those data have shown that regucalcin antibody has a correlation with cell differentiation and histological types. The frequency of antibody against regucalcin in well-differentiated (grade I-II) hepatocellular carcinoma is significantly higher than that in poorly differentiated (grade IIIV) ones. This implies regucalcin playing a role in prognosis prediction of HCC. Anti-regucalcin antibody in the HCC sera with alpha-fetoprotein negative is higher (43.6%) compared with those sera with AFP positive (26.2%), even though the positive rate for regucalcin antibody is not related to AFP concentration levels [15], see Table 2.

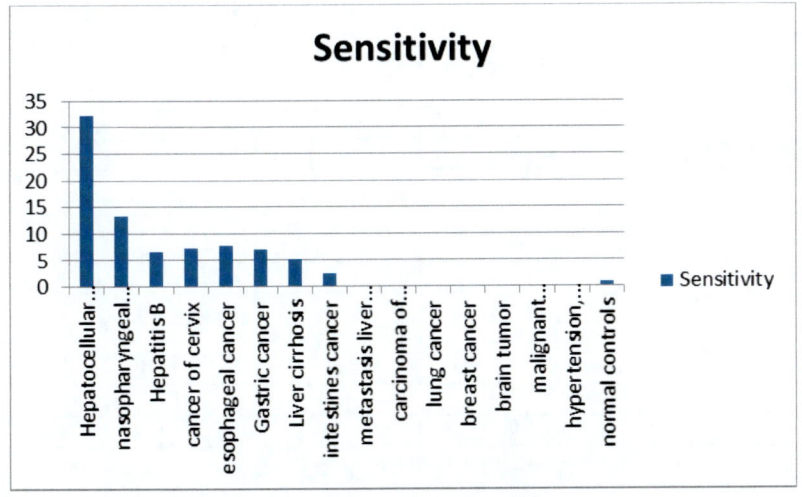

Figure 3. Showing the sensitivity of anti-regucalcin autoantibody for detecting different cases. Bar graphs show the sensitivity of each case.

Table 2. The correlations between sera antibodies to regucalcin in HCC patients and clinical parameters [15]

Clinical parameter		Number of patients (n)	Positive rate of antibodies	P value
Age	<30	11	9.1	>0.05
	30~60	108	33.3	
	>60	26	38.5	
AFP	>25ng/ml	103	26.2	<0.05
	<25ng/ml	39	43.6	
	Not available	3		
AFP level	25~400ng/ml	16	25.0	>0.05
	>400ng/ml	87	26.4	
	Not available	42		
Size of tumor	≤5	38	39.5	>0.05
	>5	95	29.5	
	Not available	12		
No of tumor	single	87	34.5	>0.05
	multiple	52	27.0	
	Not available	6		
Pathologic grade*	I~II	19	52.6	<0.05
	III~IV	21	19.0	
	Not available	105		
TNM classification	I~II	58	36.2	>0.05
	III	10	40.0	
	IV	65	29.2	
	Not available	12		
HBsAg	positive	121	31.4	>0.05
	negative	18	22.2	
	Not available	6		
Liver cirrhosis	yes	39	43.6	>0.05
	no	46	52.2	
	Not available	60		
metastasis	yes	11	36.3	>0.05
	no	124	27.4	
	Not available	10		
Portal vein embolization	yes	37	29.7	>0.05
	no	95	34.7	
	Not available	13		

*HCC pathological grade is according to 4 Edmondson grade.

The Underlying Mechanisms of the Generations of Anti-Regucalcin Antibody in HCC Patient's Serum Remain Poorly Understood

The mechanisms of underlying T-cell-mediated humoral immune response to cancer-associated antigens are still not fully understood and the subject is being intensively investigated. The production of AAbs is thought to be elicited by proteins presenting neoepitopes, for example, resulting from somatic missense mutations in gene coding domains or from post-translational modifications such as glycosylation, phosphorylation, polyadenylation, adenosine diphosphate- ribosylation, alteration of glycation side-chains. Alternatively, it has been proposed that the AAb production may also be elicited by locally aberrant expression levels of proteins. Each of these aberrations can result in detection by immune surveillance systems, activation of B- and T-lymphocytes, and the release of chemokines and cytokines, which, in turn, further stimulate the immune response. The B-lymphocytes provide a targeted antibody response to TAAs, resulting in antigen-specific AAbs.

Local inflammation and immune cell infiltration in the tumor microenvironment is a likely enhancer, if not a pre-requisite, for immune reaction and antibody formation [1, 16-18]. Serum concentration of regucalcin antigen has been detected in HCC patients at 8.39%, which is much lower than the positive rate of the serum concentration of anti-regucalcin antibodies at 32.4% in HCC patients. Researches have shown that there is almost universal down-regulation of regucalcin in HCC tissues and up-regulation in paracancerous tissues, while the autoantibody has only appeared in parts of HCC patients.

The mutation and copy number alteration data have been analyzed from TCGA liver hepatocellular carcinoma dataset using the cBioPortal. In total, genetic alterations of regucalcin have been identified in 6/440 (1.4%), with 1 in amplification, 2 in missense mutation, and 4 in deep deletion. The very low frequency of genetic alterations in regucalcin indicates that other mechanisms, such as post-translational modifications, accounts for immunogenicity to regucalcin protein in tumor patients.

REGUCALCIN: A TUMOR-SUPPRESSING MOLECULE

A tumor-suppressing gene is a gene that protects a cell from one step on the path to cancer. When this gene expression reduction or mutation causing a loss in its functionality, the cell can progress to cancer, usually in the combination with other genetic changes. The loss of these genes may be even more important than proto-oncogene/oncogene activation for the formation of many kinds of human cancer cells [19]. A large number of studies have shown that the inhibition of regucalcin gene expression involves carcinogenesis.

Regucalcin Expression Level Being Down-Regulated during Carcinogenesis

The expression levels of regucalcin protein in various human normal tissues and liver cancer tissues have been evaluated using immunohistochemical (IHC) staining. Results have shown that there is a preferentially expressed regucalcin in normal liver compared to other tissues, such as lung, spleen, myocardium, prostate and skin etc. In contrast to the higher expression levels of regucalcin protein in normal liver tissues, its expression levels in liver cancer tissues are significantly lower than that in paired adjacent non-tumor tissues. The regucalcin mRNA expression levels in paired HCC and its adjacent tissues have also been evaluated using real-time quantitative PCR and *in situ* hybridization. Results have shown that there are significant down-regulations of regucalcin mRNA in matched tumors.

Many studies carried out on multi-gene expression profiles and proteomics have also mentioned the regucalcin gene and its protein expression levels specifically to down-regulate tumor tissues of human subjects, including kidney, lung, brain, breast and prostate etc. as listed in Table 3 [20]. Three independent microarray datasets from publicly available Oncomine databases [21, 22] have shown that the mRNA expression levels of regucalcin are in consistent with the above experimental data. This suggests that the loss of regucalcin may be associated with tumor onset and progression. Regucalcin may play a role as a suppressing protein in carcinogenesis. Compared to the low expression levels of regucalcin protein in tumor tissue, the expression levels of regucalcin in tumor cell lines vary widely even with the same tumor tissue. The cell lines with different origins have very different regucalcin expression levels as summarized in Table 3 [24-27].

Table 3. The expression levels of regucalcin in human tumor tissues and cell lines [20, 24-26]

Tissue or Cell line	Tumor type	Biomolecule	Expression status
Liver tissue	Hepatocellular carcinoma	mRNA、protein	Low expression
Kidney tissue	Transitional cell carcinoma	mRNA	Low expression
Brain tissues	Meningioma	mRNA	Low expression
Lung tissues	Non-small cell carcinoma Lung squamous carcinoma	mRNA、protein	Low expression
Breast tissues	Infiltrating ductal carcinoma	mRNA、protein	Low expression
Prostate tissues	Adenocarcinoma	mRNA、protein	Low expression
GS-HepG2/ HepG2/ Hep3B cell lines	Hepatocellular carcinoma	mRNA、protein	High expression
HuH-7 cell lines	Hepatocellular carcinoma	mRNA、protein	Middle expression
SK-hep1 cell lines	Hepatocellular carcinoma	protein	Low expression
A549 cell lines	lung cancer	protein	Low expression
A498 cell lines	Renal cell carcinoma	protein	Low expression
RKO cell lines	colorectal carcinoma	protein	Low expression

Regucalcin Low-Expression Being Related to Poor Overall Survival Rate of Tumor Patients

It has been discovered that the decreased expression of regucalcin in hepatocellular carcinoma is noticeably related to larger tumor sizes and advanced TNM stage. There are no significant differences among regucalcin expression and gender, age, liver cirrhosis, histopathologic grading, capsular formation or vascular invasion. A higher regucalcin expression in HCC tissues has been demonstrated to prolong the survival of patients. The regucalcin expression and its correlations with overall survival (OS) in hepatocellular carcinoma patients have also been evaluated using histopathologic grading. The results of Kaplan-Meier analyses have suggested that the shorter the survival time of patients with HCC, the lower the expression of regucalcin. This is not related to histopathological grade, indicating that regucalcin can be used as an independent predictor of poor prognosis [23].

Data mining and analyses of HCC patients with different regucalcin mRNA levels from publicly available Oncomine databases have been carried out to confirm the prognosis significance of regucalcin in HCC. Low levels of regucalcin are found to be related to OS rate of HCC patients, suggesting that down-regulated regucalcin gene expression indicates poor prognosis of HCC patients. Other published reports have also demonstrated that survival is prolonged in patients with pancreatic cancer, breast cancer, colorectal cancer and lung cancer who have a higher regucalcin expressions in their tumor tissues as compared with those with a lower regucalcin expressions [23-29].

Regucalcin Expression is Regulated by a Variety of Factors and Its Low Expression in Tumor Tissues Is Associated with Promoter DNA Methylation

Researches have shown that the expression of regucalcin mRNA has been stimulated by hormonal factors. The myriad of factors exerts up-regulation effects (including calcium, calcitonin, parathyroid hormone, insulin, estrogen and dexamethasone) *in vivo*. Other up-regulated or down-regulated regucalcin expression (including Triiodothyronine, 5α-dihydrotestosterone, 17β-estradiol, Aldosterone) depends on the cell type, doses and/or time of stimulation. Some factors represent inhibition effects (including lipopolysaccharide, carbon tetrachloride) as illustrated in Figure 4 [20, 30]. The regucalcin gene consists of seven exons, six introns and several consensus regulatory elements exist upstream of the 50-flanking region.

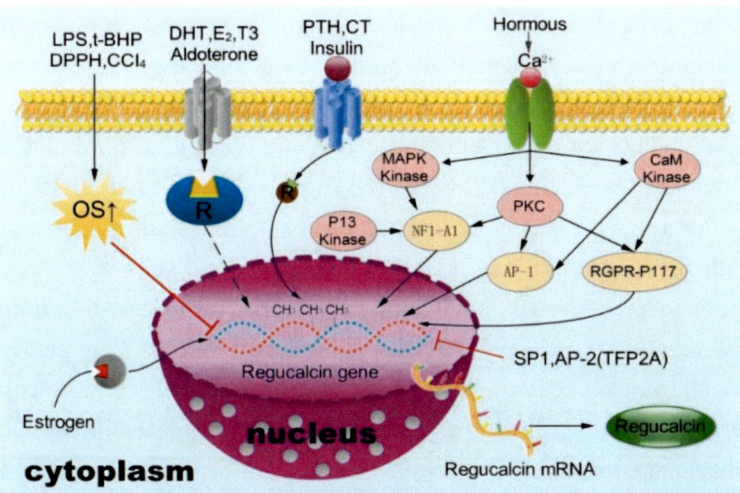

Figure 4. Illustrating that the transcription of the regucalcin gene is regulated by a variety of factors. Legend: Head-arrow lines and bar-ended lines indicate activation and inhibition, respectively. Dashed arrows indicate that up- or down-regulation may occur dependently on the dose, time of stimulation and specific cell type. LPS, lipopolysaccharide; t-BHP, tert-Butyl hydroperoxide; DPPH, 1,1-diphenyl1-2-picryl-hydrazyl; CCl4, carbon tetrachloride; OS, oxidative stress; DHT, 5α-dihydrotestosterone; R, receptor; E2, 17β-estradiol; T3, Triiodothyronine; PTH, parathyroid hormone; CT, calcitonin; CaM, calmodulin; PKC, protein kinase C [20, 32].

Studies have shown that the regulation of RGN is affected by multiple transcription factors, such as AP-1, NFI-A1, RGPRp117 and β-catenin. Those have been found to be the transcription factors to enhance the regucalcin gene promoter activity. The AP-1, NF1-A1 and RGPR-p117 of transcription factors are translocated from the cytoplasm to the nucleus. They are mediated through protein kinase C, Ca^{2+}-calmodulin-dependent protein kinase (CaM kinase), MAPK kinase and PI3 kinase, which are activated by various hormones as illustrated in Figure 4. Those transcription factors enhance the promoter activity of regucalcin gene in the nucleus. The SP1 as a transcription factor is located within -188/-180 of the promoter in HepG2 cells to suppress the regucalcin gene promoter activity [31].

It has been reported that the expression of regucalcin in the liver cancer and other cancer state has shown to be suppressed, suggesting an involvement of regucalcin in the disease. But the regulation mechanism of regucalcin expression change in human carcinogenesis remains unclear. The Oncoprint feature of the cBioPortal has been used to determine the mutation and copy number alteration frequency of regucalcin in HCC. The regucalcin has been found less altered in 6 (1.4%) of queried samples. It seems that regucalcin genetic alterations are not the main mechanism that accounts for regucalcin loss in HCC patients.

Studies have shown that the methylation of regucalcin promoter is closely related to its low expression in liver cancer. The methylation modification of regucalcin gene has been observed in three HCC cell lines by pyrosequencing. The lower CpG methylation appears in MHCC97-H and Huh7 hepatoma cell lines with high regucalcin expression and higher CpG methylation occurs in the SK-HEP-1 cell line with low regucalcin expression. The methylation levels of regucalcin in 10 pairs of matched HCC and paracancerous samples have been examined. Resluts show that HCC tissue samples have higher levels of DNA methylation at each CpG site, with corresponding hypomethylation in non-cancer tissue.

Further analyses of RNAseq data and Methylation 450k data from 371 HCC patients in UCSC Xena have observed that regucalcin expression level is also inversely proportional to the degree of DNA methylation. The regucalcin DNA methylation-mediated transcriptional silencing has been implicated as a cause of liver cancer. The additions of the histone deacetylase inhibitor, PBA, have no significant effect on the expression levels of regucalcin. This suggests that the low expression of regucalcin in HCC patients is primarily regulated by DNA methylation not by histone acetylation.

The Transcription Factor TFAP2A, SP1 Binding to the Regucalcin Gene Promoter to Reduce Its Expressions

The DNA pulldown and LC-MS methods have been used to search for and identify proteins that binding to the CpG sequence (-122 to -1bp) of the regucalcin promoter region. Results have shown that transcription factor TFAP2A binds to the CpG sequence (-122 to -1bp) of the regucalcin promoter region. It has been confirmed that the TFAP2A protein directly binds to the CpG sequence (-122 to -1bp) of the regucalcin promoter region through ChIP assays in hepatocellular carcinoma SK-hep1 cell lines. Dual luciferase assays have been used to verify the transcriptional regulatory role of TFAP2A for regucalcin gene. Results have shown that TFAP2A significantly inhibits the activity of wild-type promoter (-522 to -1bp) of regucalcin. But it does not obviously inhibit the activity of truncated regucalcin promoter (-522 to -123bp) lacking CpG sequence (-1 to -122 bp). Those results indicate that the binding site of TFAP2A is at the CpG sequence (-122 to -1bp) of the regucalcin promoter region. After the knockdown of TFAP2A by siRNA, regucalcin expression is found to be elevated.

The influence of regucalcin promoter CpG sequence methylation on the binding of transcription factors TFAP2A has been evaluated. The ChIP assays have been performed using SK-HEP-1cell lines that have a CpG sequence of hypermethylation in the promoter and low regucalcin expression. The results show that the amount of TFAP2A protein bound to the CpG sequence (-122 to -1 bp) of regucalcin promoter region has decreased significantly with the addition of the DNA methyltransferase inhibitor DAC. The amount of TFAP2A bound to regucalcin promoter region has reduced when DAC is added. This suggests that the TFAP2A can easily bind to methylation regucalcin promoter. The above results indicate that methylation of the CpG sequence (-122 to -1bp) of the regucalcin promoter region promotes the binding of TFAP2A and inhibits regucalcin expression as illustrated in Figure 5.

Another research also reports that transcription factor SP1 binds to the CpG sequence of the regucalcin promoter region to decrease regucalcin expression. Regucalcin expression has shown to be up-regulated after phorbol 12-myristate 13-acetate (PMA) treatment. Pretreatment with CHX, a protein synthesis inhibitor, does not significantly inhibit the effect of PMA-induced up-regulation of RGN as analyzed by real-time PCR, suggesting that human RGN is up-regulated by PMA treatment independent of translation. The SP1 dissociates from a SP1 motif located within -188/-180 in the promoter after PMA induction, which thus leads to the activation of both internal expression and promoter activities in HepG2 cells. This is the mechanism by which PMA up-regulates the expression of human RGN via driving SP1 away from a SP1 motif located within -188/180 of the promoter. Overexpression of SP1 dramatically reduces PMA-induced up-regulation of both internal expression of mRNA and promoter activity. The knockdown of SP1 has the opposite effect [31].

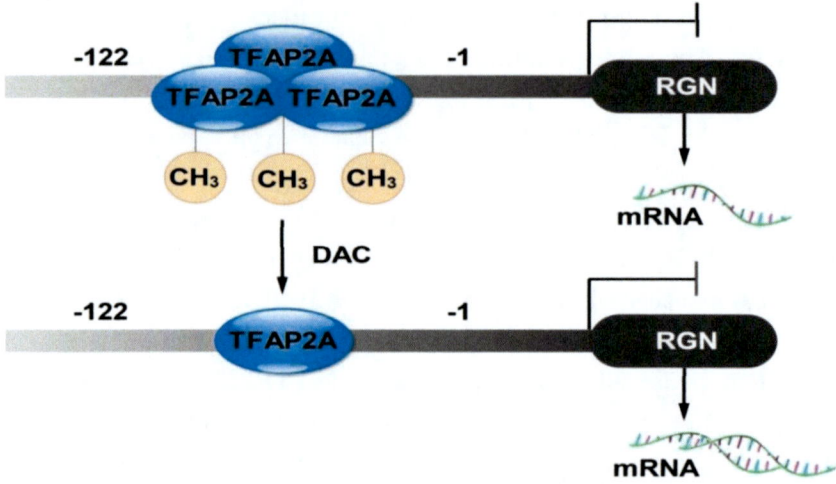

Figure 5. Illustrating the transcription factor TFAP2A bound to the CpG sequence (-122 to -1 bp) of regucalcin (RGN) promoter region decreased by DNA methyltransferase inhibitor DAC. Methylation of the CpG sequence (-122 to -1bp) of the regucalcin promoter region promotes the binding of TFAP2A and inhibits regucalcin expression. The amount of TFAP2A protein bound to the CpG sequence (-122 to -1 bp) of the promoter region is decreased significantly as the degree of methylation decreased, after the addition of the DNA methyltransferase inhibitor DAC.

Downregulation of Regucalcin Enhancing Tumor Cell Migration and Invasion through EMT and Wnt/B-Catenin Signaling Pathways

In vitro with overexpression of regucalcin in human hepatocellular carcinoma HepG2 cells, lung adenocarcinoma NSCLC A549 cells, breast cancer MDA-MB-231cells, pancreatic cancer MIA PaCa-2 cells and cervical cancer HeLa cells, the migration and invasion of tumor cells are suppressed [26-29]. Lentivirus-mediated regucalcin-shRNA is transfected into HeLa cells, which accelerate the invasion and migration of cells. Mechanistically, it has demonstrated that the expression of E-cadherin, a classical marker of mesenchymal phenotype, is decreased after the down-expression of regucalcin in HeLa cells. The expression of vimentin and N-cadherin is increased, indicating that the EMT process has been activated by the down-expression of regucalcin. This promotes the invasion and migration of tumor cells.

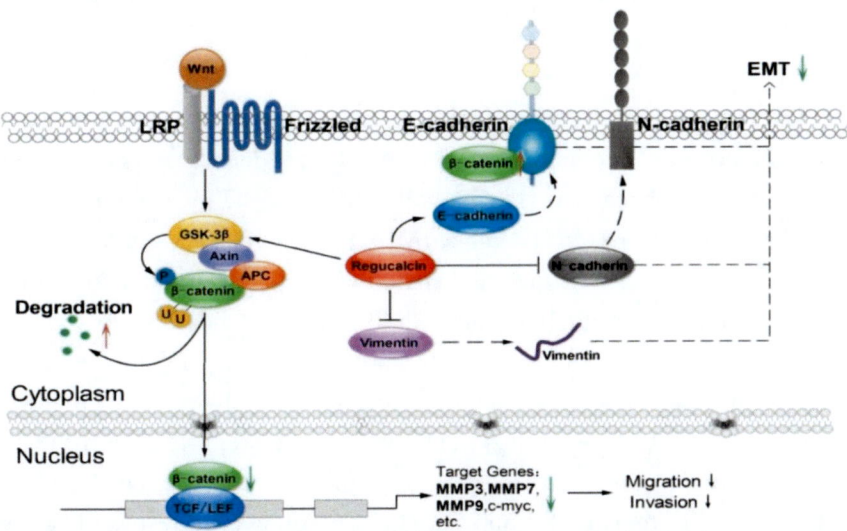

Figure 6. Illustrating the mechanisms by which regucalcin (RGN) controls intracellular signaling, cell EMT, migration and invasion. The RGN increases E-Cadherin expression levels and inhibits the expression of Vimentin and N-Cadherin, which leads to the inhibition of cell EMT process. The RGN has the ability to enhance GSK-3β activity, which leads to much β-catenin protein degraded by ubiquitination and rare β-catenin entering the nucleus. The results eventually diminish the expression levels of target genes (such as MMP3\MMP7\MMP9), the transcription and expression of downstream genes such as MMP3\MMP7\MMP9, which leads to the invasion and migration of tumor cells. Legend: Head-arrow lines and bar-ended lines indicate activation and inhibition, respectively. Dashed arrows indicate that up- or down-regulation of EMT occur dependently on the cell microenvironment [26-29, 33].

Through the detection of key genes in the Wnt/β-catenin pathway, it has been found that regucalcin down-regulation lead to the increase of p-GSK-3β expression. This means that GSK-3β activity has reduced and less β-catenin has been degraded by ubiquitination. This results in excessive β-catenin entering the nucleus, forming more β-catenin/LEF/TCF complex, promoting the transcription and expression of downstream genes such as MMP3\MMP7\MMP9. It also promotes the invasion and migration of HeLa cells as illustrated in Figure 6 [33].

Regucalcin Inhibiting Tumor Cell Proliferation Independent of Cell Death

Overexpression of regucalcin decreases the proliferation index in human colorectal carcinoma RKO cells, lung adenocarcinoma NSCLC A549 cells, liver cancer HepG2 cells, MDA-MB-231 bone metastatic breast cancer cells, pancreatic cancer MIA PaCa-2 cells and cervical cancer HeLa cells, and also in the rat hepatoma H4-II-E cells and rat kidney proximal tubular epithelial cell [25-29, 33-35]. The mechanism of regucalcin inhibits tumor cell proliferation is mainly by the enhanced expression of several tumor suppressing genes (namely, p21, p53 and Rb), suppressed levels of oncogenes (namely, H-ras, c-src, c-fos, c-myc, c-Jun and c-kit) and counteracted cell cycle regulators (namely, cdc2, chk2m, cdk 5, and SCF) to cause G1 and G2/M phase cell cycle arrest as illustrated in Figure 7 [34-36].

It has been proposed that regucalcin suppresses cell proliferation *in vitro* by reducing protein kinase activity and protein phosphatase activity in cytoplasm and nuclei, such as Ca^{2+}/calmodulin-dependent protein kinase(CaMK), protein kinase C (PKC), protein tyrosine kinase(PTK), mitogen-activated protein kinase(MAPK), phosphatidylinositol 3-kinase (PI3K), extracellular signal-regulated kinase(ERK), Ca^{2+}/calmodulin-dependent protein tyrosine phosphatase((CaMKP) and protein tyrosine phosphatase (PTP) etc [34-36].

The overexpression of regucalcin suppresses the death of human colorectal carcinoma RKO cells, human hepatoma HepG2 cells, lung adenocarcinoma NSCLC A549 cells and human clear cell renal cell carcinoma (RCC) A498 cells, cultured in a medium containing fetal bovine serum with various factor known to induce apoptotic cell death *in vitro*. Regucalcin increases cell survival in response to Ca^{2+} ionophore A23187. This induces intracellular Ca^{2+} overcharge, thapsigargin, actinomycin D, H_2O_2, t-BHP, insulin, transforming growth factor (TGF)-β, tumor necrosis factor (TNF)-α, insulin-like growth factor or LPS. All of those are known inducers of oxidative stress, DNA fragmentation and cell death.

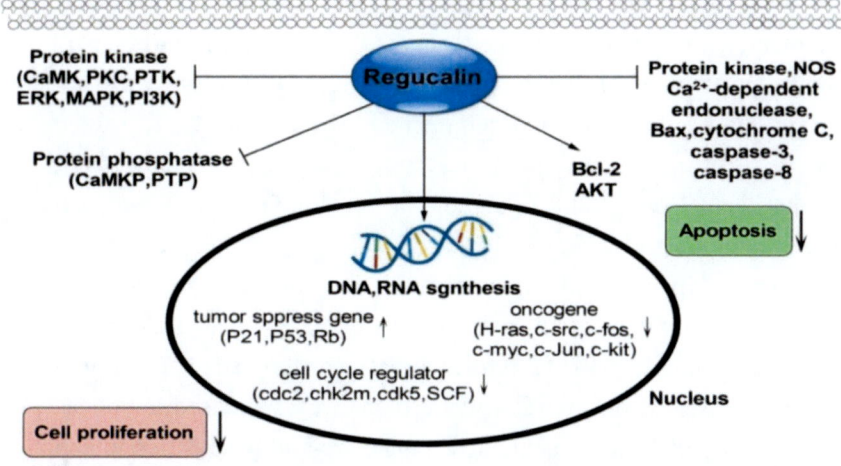

Figure 7. Illustrating regucalcin affecting cell proliferation and apoptosis by regulating intracellular signaling. Regucalcin inhibits activities of protein kinase, protein phosphatase in cytoplasm and nuclei. In addition, RGN enhances the expression levels of several tumor suppressing genes and decreases the expression levels of oncogenes, which leads to the inhibition of cell proliferation. The inhibitory effects of RGN on apoptosis is related to the regulation of the expression and activity of various apoptosis-related proteins, such as activation of anti-apoptotic proteins Bcl-2 and AKT and inhibition of apoptosis-inducing proteins Bax, cytochrome C, caspase-3, caspase-8, protein kinases, NO synthase and Ca2+-dependent endonuclease [32, 37].

The above suggests that the suppressive effects of regucalcin overexpression on cell growth do not result in the death of tumor cells. The inhibitory effect of regucalcin on apoptosis may be related to its suppressive effect on the expression and activity of various protein kinases, nitric oxide synthase (NOS), Ca^{2+}-dependent endonuclease, Bax, cytochrome C and caspase-3 and also with the stimulation of anti-apoptotic protein Bcl-2 as illustrated in Figure 7 [37].

All of those findings are confirmative of the involvement of regucalcin as a protein that protects from cell death induced by damaging factors, thus, contributing to the maintenance of healthy tissues. Regucalcin also seems to be able to modulate cell death in NRK52E and HepG2 cells and also in the prostate and mammary gland tissues, even without any treatment with apoptosis-inducers agents. The NRK52E and HepG2 cells transfected with regucalcin have shown increased expression of Bcl-2 and protein kinase B(AKT) and augmented activity of AKT, respectively [38,39]. Further research is warranted to deeply characterize the peculiarities of apoptotic regulation by regucalcin in different cells and tissues contexts.

Several types of experimental evidence have demonstrated that the suppression of regucalcin expression increases cell susceptibility to apoptotic death. In MCF-7 human breast cancer cells the repression of regucalcin expression by T3 is underpinned by increased apoptosis [40]. The hepatocytes from regucalcin-KO mice are highly sensitive to apoptosis induced by TNF-α plus actinomycin D, which seems to involve the activation of caspase-8 but not the NF-kB and Fas-mediated apoptosis [41]. The RGN generally acts as an anti-apoptotic protein. But in aged animals it has been found that RGN overexpression favors apoptosis [42], which may be a protective mechanism against age-associated diseases.

On a whole, regucalcin plays a crucial role as a suppressor in human cancer. The suppressed expression of the regucalcin gene may predispose patients to the promotion of cancer. The overexpression of regucalcin by gene delivery may thus prove to be a novel therapeutic strategy for cancer.

Conclusion

Regucalcin mRNA and its protein content are mainly expressed in the liver and kidney cortex, beside the expression in other tissues and cell types. Regucalcin is generally under-expressed in tumor tissues. However, anti-regucalcin antibody can be detected in the blood of some tumor patients, such as liver cancer and gastric cancer. The anti-regucalcin antibody positive rate is higher in AFP-negative liver cancer serum than that in AFP-positive liver cancer serum. The frequency of serum antibody against regucalcin with well histopathologically differentiated samples is significantly higher than that in poorly differentiated ones.

The mechanism for producing anti-regucalcin antibodies remains unclear. But it is probably related to the post-translational modification of proteins. The low expression of regucalcin in tumor tissues is associated with shorter survival time of patients. The expression of regucalcin gene is regulated through various hormone, transcription factors and epigenetic alterations. The decreased expression of regucalcin in liver cancer is inversely proportional to the promoter methylation frequency. The transcription factor TFAP2A binds to the promoter CpG sequence (-122 to -1bp) to inhibit the expression of regucalcin.

The experiments *in vitro* have shown that regucalcin inhibits tumorigenesis through a variety of pathways. Regucalcin inhibits the proliferation of tumor cells by increasing the expression of tumor suppressing genes such as P53 and P21. It also promotes apoptosis of tumor cells by inhibiting the expression of oncogenes c-jun, H-ras and myc, etc. It has been proposed that regucalcin suppresses cell proliferation *in vitro* by reducing protein kinase activity and protein phosphatase activity in cytoplasm and nuclei. Regucalcin can inhibit the invasion and migration of tumor cells by inhibiting Wnt/β-catenin signaling pathway. The suppressive effects of regucalcin overexpression on cell growth are not resulted from the death of tumor cells.

REFERENCES

[1] Zaenker P, Gray ES, Ziman MR (2016) Autoantibody production in cancer - the humoral immune response toward autologous antigens in cancer patients, *Autoimmun Rev* 15:477-483.

[2] Balan S, Finnigan J, Bhardwaj N (2017) Dendritic cell strategies for eliciting mutation-derived tumor antigen responses in patients. *Cancer J* 23:131-137.

[3] Lu H, Goodell V, Disis ML (2008) Humoral immunity directed against tumor-associated antigens as potential biomarkers for the early diagnosis of cancer. *J Proteome Res* 7:1388-1394.

[4] Anderson KS, LaBaer J (2005) The sentinel within: exploiting the immune system for cancer biomarkers. *J Proteome Res* 4:1123–1133.

[5] Zayakin P, Ancāns G, Siliņa K, Meistere I, Kalniņa Z, Andrejeva D, Endzeliņš E, Ivanova L, Pismennaja A, Ruskule A, Doniņa S, Wex T, Malfertheiner P, Leja M, Linē A (2013) Tumor-associated autoantibody signature for the early detection of gastric cancer. *Int J Cancer* 132137-147.

[6] Zaenker P, Ziman MR (2013) Serologic autoantibodies as diagnostic cancer biomarkers--a review. *Cancer Epidemiol Biomarkers Prev* 22:2161-181.

[7] Sahin U, Türeci O, Schmitt H, Cochlovius B, Johannes T, Schmits R, Stenner F, Luo G, Schobert I, Pfreundschuh M (1995) Human neoplasms elicit multiple specific immune responses in the autologous host. *Proc Natl Acad Sci USA* 92:11810-1193.

[8] Sahin U, Türeci O, Pfreundschuh M (1997) Serological identification of human tumor antigens. *Curr Opin Immunol* 9:709-716.

[9] Stenner-Liewen F, Luo G, Sahin U, et al. (2000) Definition of tumor associated antigens in hepatocellular carcinoma. *Cancer Epidemiol Biomarkers Prev* 9:285-290.

[10] Zhou SF, Xie XX, Bin YH, Lan L, Chen F, Luo GR (2006) Identification of HCC-22-5 tumor-associated antigen and antibody response in patients. *Clinica Chimica Acta* 366: 274-280.

[11] Fujita T (1999) Senescence marker protein-30 (SMP30): structure and biological function. *Biochem Biophys Res Commun* 254:1-4.
[12] Ishigami T, Fujita T, Simbula G, Columbano A, Kikuchi K, Ishigami A, Shimosawa T, Arakawa Y, Maruyama N (2001) Regulatory effects of senescence marker protein 30 on the proliferation of hepatocytes. *Pathol Int* 51:491-497.
[13] Ushigome M, Nabeya Y, Soda H, Takiguchi N, Kuwajima A, Tagawa M, Matsushita K, Koike J, Funahashi K, Shimada H (2018) Multi-panel assay of serum autoantibodies in colorectal cancer. *Int J Clin Oncol* 23:917-923.
[14] Fujita T, Uchida K, Maruyama N (1992) Purification of senescence marker protein-30 (SMP30) and its androgen-independent decrease with age in the rat liver. *Biochim Biophys Acta* 1116:122-128.
[15] Zhou SF, Mo FR, Bin YH, Hou GQ, Xie XX, Luo GR (2011) Serum immunoreactivity of SMP30 and its tissues expression in hepatocellular carcinoma. *Clin Biochem* 44:331-336.
[16] Hoshino I, Nagata M, Takiguchi N, Nabeya Y, Ikeda A, Yokoi S, Kuwajima A, Tagawa M, Matsushita K, Satoshi Y, Hideaki S (2017) Panel of autoantibodies against multiple tumor-associated antigens for detecting gastric cancer. *Cancer Sci* 108:308-315.
[17] Khodadoust MS, Alizadeh AA (2014) Tumor antigen discovery through translation of the cancer genome. *Immunol Res* 58:292-299.
[18] Fortner RT, Damms-Machado A, Kaaks R (2017) Systematic review: Tumor-associated antigen autoantibodies and ovarian cancer early detection. *Gynecol Oncol* 147:465-480.
[19] Robert A .Weinberg. *The Biology of Cancer* (Second Edition). Garland Science, 2013, page 231.
[20] Vaz CV, Correia S, Cardoso HJ, Figueira MI, Marques R, Maia CJ, Socorro S (2016) The Emerging Role of Regucalcin as a Tumor Suppressor: Facts and Views. *Curr Mol Med* 16:607-619.
[21] Chen X, Cheung ST, So S, Fan ST, Barry C, Higgins J, Lai KM, Ji J, Dudoit S, Ng IO, Van De Rijn M, Botstein D, Brown PO (2002) Gene expression patterns in human liver cancers. *Mol Biol Cell* 13:1929-1939.

[22] Roessler S, Jia HL, Budhu A, Forgues M, Ye QH, Lee JS, Thorgeirsson SS, Sun Z, Tang ZY, Qin LX, Wang XW (2010) A unique metastasis gene signature enables prediction of tumor relapse in early-stage hepatocellular carcinoma patients. *Cancer Res* 70:10202-10212.

[23] Mo Z, Zheng S, Lv Z, Zhuang Y, Lan X, Wang F, Lu X, Zhao Y, Zhou S (2016) Senescence marker protein 30 (SMP30) serves as a potential prognostic indicator in hepatocellular carcinoma. *Scientific Reports* 6:39376.

[24] Yamaguchi M, Osuka S, Hankinson O, Murata T (2019) Prolonged survival of renal cancer patients is concomitant with a higher regucalcin gene expression in tumor tissues: Overexpression of regucalcin suppresses the growth of human renal cell carcinoma cells *in vitro*. *Int J Oncol* 54:188-198.

[25] Yamaguchi M, Osuka S, Murata T (2018) Prolonged survival of patients with colorectal cancer is associated with a higher regucalcin gene expression: Overexpression of regucalcin suppresses the growth of human colorectal carcinoma cells *in vitro*. *Int J Oncol* 53:1313-1322.

[26] Yamaguchi M, Osuka S, Shoji M, Weitzmann MN, Murata T (2017) Survival of lung cancer patients is prolonged with higher regucalcin gene expression: suppressed proliferation of lung adenocarcinoma A549 cells *in vitro*. *Mol Cell Biochem* 430:37-46.

[27] Yamaguchi M, Osuka S, Weitzmann MN, El-Rayes BF, Shoji M, Murata T (2016) Prolonged survival in hepatocarcinoma patients with increased regucalcin gene expression: HepG2 cell proliferation is suppressed by overexpression of regucalcin *in vitro*. *Int J Oncol* 49:1686-1694.

[28] Yamaguchi M, Osuka S, Weitzmann MN, Shoji M, Murata T (2016) Increased regucalcin gene expression extends survival in breast cancer patients: Overexpression of regucalcin suppresses the proliferation and metastatic bone activity in MDA-MB-231 human breast cancer cells *in vitro*. *Int J Oncol* 49:812-822.

[29] Yamaguchi M, Osuka S, Weitzmann MN, El-Rayes BF, Shoji M, Murata T (2016) Prolonged survival in pancreatic cancer patients with increased regucalcin gene expression: Overexpression of regucalcin suppresses the proliferation in human pancreatic cancer MIA PaCa-2 cells *in vitro*. *Int J Oncol* 48:1955-1964.

[30] Yamaguchi M (2011) The transcriptional regulation of regucalcin gene expression. *Mol Cell Biochem* 346:147-1471.

[31] Xu H, Ni P, Chen C, Yao Y, Zhao X, Qian G, Fan X, Ge S (2011) SP1 suppresses phorbol 12-myristate 13-acetate induced up-regulation of human regucalcin expression in liver cancer cells. *Mol Cell Biochem* 355:9-15.

[32] Yamaguchi M (2013) Suppressive role of regucalcin in liver cell proliferation: involvement in carcinogenesis. *Cell Prolife* 46: 243–253.

[33] Li XL, Huang YW, Wang PF, Song W, Yao QM, Hu QP, Zhou SF (2019) A mechanism of regucalcin knock-down in the promotion of proliferation and movement of human cervical cancer HeLa cells. *Transl Cancer Res* doi: 10.21037/tcr.2019.02.01.

[34] Yamaguchi M, Daimon Y (2005) Overexpression of regucalcin suppresses cell proliferation in cloned rat hepatoma H4-II-E cells: Involvement of intracellular signaling factors and cell cycle-related genes. *J Cell Biochem* 95: 1169-1177.

[35] Nakagawa T, Sawada N, Yamaguchi M (2005) Overexpression of regucalcin suppresses cell proliferation of cloned normal rat kidney proximal tubular epithelial NRK52E cells. *Int J Mol Med* 16: 637-643.

[36] Yamaguchi M, Murata T (2015) Suppressive effects of exogenous regucalcin on the proliferation of human pancreatic cancer MIA PaCa-2 cells *in vitro*. *Int J Mol Med* 35:1773-1778.

[37] Yamaguchi M (2013) The anti-apoptotic effect of regucalcin is mediated through multisignaling pathways. *Apoptosis* 18:1145-1153.

[38] Nakagawa T, Yamaguchi M (2005) Overexpression of regucalcin suppresses apoptotic cell death in cloned normal rat kidney proximal

tubular epithelial NRK52E cells: Change in apoptosis-related gene expression. *J Cell Biochem* 96:1274-1285.

[39] Matsuyama S, Kitamura T, Enomoto N, Fujita T, Ishigami A, Handa S, Maruyama N, Zheng D, Ikejima K, Takei Y, Sato N (2004) Senescence marker protein-30 regulates Akt activity and contributes to cell survival in Hep G2 cells. *Biochem Biophys Res Commun* 321:386-390.

[40] Sar P, Peter R, Rath B, Das Mohapatra A, Mishra SK (2011) 3, 3', 5 Triiodo L thyronine induces apoptosis in human breast cancer MCF-7 cells, repressing SMP30 expression through negative thyroid response elements. *PLoS One* 6: e20861.

[41] Ishigami A, Fujita T, Handa S, Shirasawa T, Koseki H, Kitamura T, Enomoto N, Sato N, Shimosawa T, Maruyama N (2002) Senescence marker protein-30 knockout mouse liver is highly susceptible to tumor necrosis factor-alpha- and Fas-mediated apoptosis. *Am J Pathol* 161:1273-1281.

[42] Vaz CV, Marques R, Maia CJ, Socorro S (2015) Aging-associated changes in oxidative stress, cell proliferation, and apoptosis are prevented in the prostate of transgenic rats overexpressing regucalcin. *Transl Res* 166:693-705.

In: Regucalcin
Editor: Masayoshi Yamaguchi

ISBN: 978-1-53616-172-4
© 2019 Nova Science Publishers, Inc.

Chapter 4

THE ROLE OF REGUCALCIN IN BONE REMODELING AND OSTEOPOROSIS

*Masayoshi Yamaguchi**
Department of Pathology and Laboratory Medicine,
David Geffen School of Medicine, University of California,
Los Angeles (UCLA), Los Angeles, CA, US

ABSTRACT

Regucalcin was discovered as a calcium-binding protein in1978, and this protein has been demonstrated to play a multifunctional role in the regulation of the function of various types of cells and tissues. Regucalcin plays a pivotal role in the regulation of intracellular calcium homeostasis, various enzymes activity, cell signal transduction, nuclear function and gene expression, and cell proliferation and apoptosis. Moreover, regucalcin is found to play a role in the regulation of bone homeostasis, which is regulated by the functions of osteoclast and osteoblasts. Overexpressed regucalcin induces bone loss in regucalcin transgenic rats *in vivo* and deficiency causes osteomalacia *in vivo*. Regucalcin mRNA and its protein are expressed in the femoral tissues, bone marrow cells, and osteoblastic cells. Extracellular regucalcin stimulates

---

* Corresponding Author's E-mail: yamamasa11555@yahoo.co.jp.

osteoclastogenesis in bone marrow culture *in vitro* and *in vitro* and exhibits suppressive effects on the differentiation and mineralization in osteoblastic cells. Notably, extracellular regucalcin was found to suppress osteoblastogenesis and stimulate adipogenesis in bone marrow culture system *in vitro*. Regucalcin reveals enhancing effects on activation of nuclear factor-kapper B (NF-κB), which is mediated through tumor necrosis factor-α or the receptor activator of NF-κB ligand in preosteoblastic cells and preosteoclastic cells. Regucalcin plays a pivotal role in the regulation of bone homeostasis as a suppressor in osteoblastogenesis and an enhancer in osteoclastogenesis, demonstrating a role as a cytokine.

Keywords: regucalcin, bone remodeling, bone formation, bone resorption, osteoporosis

INTRODUCTION

Bone is a dynamic tissue that undergoes continual adaptation during vertebrate life to attain and preserve skeletal size, shape, and structural integrity, and this tissue plays a pivotal role to regulate mineral homeostasis in living body. Bone remodeling, which maintains bone mass, is artfully regulated through osteoclasts, osteoblasts and osteocytes, which are major cells in bone tissues. Many hormones, cytokines and immune systems in bone marrow microenvironment is involved in the regulation of bone homeostasis [1-4].

Decrease in bone mass with increasing age leads to osteoporosis. In recent years, there is accumulating evidence that regucalcin plays a crucial role in the regulation of bone homeostasis as an enhancer in osteoclastogenesis and a suppressor in osteoblastogenesis, demonstrating its role as a cytokine [4-7]. Notably, osteoporosis with bone loss is shown in regucalcin transgenic rats *in vivo* [7, 8]. Regucalcin may play a physiologic role in the regulation of bone metabolism.

This chapter will discuss recent topics regarding the role of regucalcin in bone remodeling and the involvement in osteoporosis.

BONE REMODELING

Bone contains over 98% of total body calcium. Bone metabolism is regulated by the functions of osteoclasts and osteoblasts, which are major cells in bone tissue [1-6]. Bone formation in adult cancellous bone takes place only at sites of bone remodeling. During this process, old bone is replaced with new at discrete sites by the basic multicellular unit, which comprises team of osteoclasts, osteoblasts and osteocytes. Bone formation occurs on periosteal surfaces by a process called modeling. In the physiologic process of bone turnover, a resorptive stimulus firstly triggers recruitment of osteoclasts to a site on bone surface. This is followed by active resorption by osteoclasts, after which cells withdraw from the bone surface and mononuclear phagocytic cells appear on the newly resorbed surface. These cells are followed by young osteoblasts, which begin the bone formation phase. After the resorbed lacunar pit is filled with new osteoid, osteoblasts become flatter and less active, with the final newly remodeled bone surface lined by flat lining cells.

Remodeling of cancellous bone begins with the retraction of lining cells that cover the bone surface [1, 4]. Osteoclasts, which develop from hematopoietic progenitors, are recruited to the site and excavate the calcified matrix. Then, the cavity is refilled by osteoblasts via a process that occurs in three distinct phases, including initiation, progression and termination. During initiation phase, osteoblasts arising from local mesenchymal stem cells assembles at the bottom of the cavity, and bone formation begins. Bone formation terminates when the cavity has been refilled, at which time the few osteoblasts that remain become the flat lining cells that cover the quiescent surfaces of bone. As bone formation progresses, some osteoblasts are entombed within the matrix as osteocytes but the majority dies by apoptosis. Once new bone formed, few osteocytes die. Viability of osteocytes is likely maintained by physiological levels of mechanical stimulation. When mechanical forces are reduced, for example in weightlessness, osteocytes die by apoptosis. This event acts as a beacon for osteoclast recruitment and generation of a new basic multicellular unit, which in turn replaces the old bone containing dead osteocytes with new bone containing viable osteocytes.

The process of bone remodeling makes bone unique among organs and tissues and adds so many levels of complexity with respect to: interactions along the remodeling sequence by systemic influences (hormones); stress action on trabecular and cortical systems (physical activity/weight bearing); growth factors produced by the bone cells which act locally on their own cell types and on the other bone cell types, or which are bound into the newly formed or resorbed bone matrix itself, or factors that come from nearby cells present in the marrow tissues.

Bone acts as major storage site for growth factors [3]. Growth factors, which are produced by osteoblasts, diffuse into newly deposited osteoid and are stored in the bone matrix including insulin-like growth factors (IGF- I and II), transforming growth factor-β1 (TGF-β1), or platelet-derived growth factor. These bone-derived factors, which can be liberated during subsequent periods of bone resorption, act in an autocrine, paracrine, or delayed paracrine fashion in the local microenvironment of the bone surface. Bone homeostasis is regulated via various factors.

REGUCALCIN STIMULATES OSTEOCLASTOGENESIS

Bone homeostasis is dexterously regulated through osteoclasts and osteoblasts in bone tissues to maintain bone mass [4]. Osteoclasts are differentiated from hematopoietic progenitors to stimulate bone resorption, and their differentiation is regulated through various bone-resorbing factors including parathyroid hormone (PTH), calcitonin and 1, 25-vitamin D_3 [1, 25(OH)$_2$D$_3$] and cytokines, including tumor necrosis factor-α (TNF-α), the receptor activator of nuclear factor kappa B ligand (RANKL) and other factors [5, 6]. The regucalcin gene is demonstrated to express in the bone marrow cells of wild type and regucalcin transgenic rats [7]. Regucalcin levels in the bone marrow cells were found to elevate in regucalcin transgenic rats with increasing age [7]. Notably, regucalcin is shown to stimulate osteoclastic bone resorption, and bone loss is induced in regucalcin-overexpressing transgenic rats *in vivo* [8-10].

Regucalcin is shown to stimulate osteoclastic bone resorption *in vitro* [10, 11]. Rat femoral-diaphyseal or -metaphyseal tissues were cultured for 48 hours in the presence of regucalcin (10^{-10} - 10^{-8} M) [10, 11], which is in the range of physiologic concentration of regucalcin in the serum of normal rats [10, 11]. Calcium content in the femoral-diaphyseal or -metaphyseal tissues was decreased by the addition of regucalcin in culture medium *in vitro* [10, 11]. The consumption of glucose and production of lactic acid, which are implicated to bone resorption stimulated by osteoclasts, in the femoral-diaphyseal or -metaphyseal tissues was enhanced by the addition of regucalcin into culture medium [10, 11]. Such effects were also observed in the presence of PTH, which stimulates bone resorption implicated to osteoclast activation [10, 11]. Thus, regucalcin exhibits a direct stimulatory effect on bone resorption in rat bone tissues *in vitro* [10, 11]. Cellular mechanically, regucalcin has been demonstrated to stimulate osteoclast-like cell formation in mouse bone marrow culture *in vitro*, and this effect was shown to suppress in the presence of various anti-bone resorbing factors, including calcitonin or 17β-estradiol, specific inhibitors of osteoclastic bone resorption [11]. Extracellular regucalcin plays a crucial role as a novel stimulatory factor in osteoclastic bone resorption.

RANKL is a pivotal cytokine to play a pivotal role in the differentiation from preosteoclast to osteoclasts in the presence of macrophage colony-stimulating factor (M-CSF) [5, 6, 12]. RANKL is produced from osteoblasts that are enhanced via action of various bone-resorbing factors, including PTH and 1, $25(OH)_2D_3$ [4-6]. RANKL binds to RANK (receptor for RANKL) in preosteoclasts and promotes differentiation to mature osteoclasts [4-6]. Bone resorption was enhanced in regucalcin transgenic rats *in vivo* [7, 9]. Regucalcin, which is expressed in preosteoclastic cells, stimulates osteoclast differentiation. Regucalcin plays a crucial role as a suppressor in the process of intracellular signaling in other tissues (including liver, kidney, heart and brain of rats) [13, 14]. The process of cellular signaling of RANKL to promote osteoclastogenesis may be stimulated via the process of extracellular signaling linked to the action of regucalcin in preosteoclastic cells.

PTH, $1,25(OH)_2D_3$ or prostaglandin E_2 are bone-resorbing factors, and these factors stimulate the expression of RANKL in osteoblasts [5, 6]. Secreted RANKL binds to RANK on preosteoclasts and promotes differentiation to osteoclasts [5, 6]. Regucalcin was found to stimulate the differentiation to osteoclastic cells of mouse bone marrow cells cultured for 7 days in the presence of both RANKL and M-CSF [11]. This potential effect revealed at the early stage of osteoclastic cell differentiation in culture with both RANKL and M-CSF [11]. Of note, regucalcin was shown to stimulate osteoclast-like cell formation in the absence of both RANKL and M-CSF [11]. Thus, regucalcin was demonstrated to stimulate the process of differentiation from preosteoclasts to osteoclasts independent on M-CSF and RANKL.

Stimulatory effects of regucalcin on osteoclastogenesis in bone marrow culture may be not involved in either enhanced expressions of RANKL in osteoblastic cells or suppressed expression of osteoprotegerin (OPG), which inhibits the binding of RANKL to RANK receptor in preosteoclasts [5, 6]. Regucalcin may stimulate osteoclastogenesis through mechanism, which is independent on RANKL and OPG. Moreover, the stimulatory effects of various bone-resorbing factors, including PTH, $1,25(OH)_2D_3$, prostaglandin E_2 or TNF-α, on osteoclastogenesis in mouse bone marrow cell culture was enhanced by extracellular regucalcin in mouse bone marrow culture *in vitro* [11]. Stimulatory effects of TNF-α on osteoclastogenesis were markedly enhanced by extracellular regucalcin [11]. TNF-α, an autocrine factor, promotes differentiation from preosteoclasts to osteoclasts, which are mediated through action of RANKL [12]. RANKL stimulates the formation of osteoclast, which is mediated via the activation of nuclear factor-kapper B (NF-κB) signaling in preosteoclasts (monocyte RAW267.4 cells). Regucalcin enhanced RANKL-induced activation of NF-κB signaling in RAW267.4 cells and promoted RANKL-stimulated osteoclastogenesis [15]. In addition, the stimulatory effects of regucalcin on osteoclastogenesis were involved in newly synthesized protein components in bone marrow cell culture system *in vitro* [15]. Extracellular regucalcin may bind to RANK or other receptors on preosteoclastic cells and stimulates the process of signal transduction related to gene expression in the cells.

PTH or 1,25(OH)$_2$D$_3$ stimulates the production of RANKL in osteoblasts [4-6]. Regucalcin enhanced stimulatory effects of PTH or 1,25(OH)$_2$D$_3$ on osteoclastic bone resorption. Stimulatory effects of PTH or 1,25(OH)$_2$D$_3$ on osteoclastic bone resorption may be partly mediated through the action of regucalcin in osteoclastogenesis and osteoclasts. Extracellular regucalcin, which directly stimulates osteoclastic bone resorption *in vitro*, may play a role as a novel bone-resorbing factor.

In vivo studies, overexpressed regucalcin was demonstrated to stimulate bone resorption in the bone tissues of regucalcin transgenic rats [8, 9]. Culture of bone marrow cells obtained from regucalcin transgenic rats produced osteoclastic cells [9, 10]. Tartrate-resistant acid phosphatase (TRACP) is a marker enzyme of osteoclasts [10]. Number of TRACP positive multinucleated cells (MNCs) was increased by the culture of bone marrow cells obtained from regucalcin transgenic rats, once bone marrow cells obtained from wild-type rats or regucalcin transgenic rats were cultured for 7 days without the addition of bone-resorbing factors *in vitro* [9]. This increase was a remarkable in regucalcin transgenic female rats as compared with that of male rats [9], indicating that the bone loss induced in regucalcin transgenic rats was remarkable in female rats as compared with that of male rats [8]. Thus, overexpressed regucalcin has been shown to stimulate osteoclast-like cell formation in bone marrow cells *in vivo*.

Notably, osteoclastic bone resorption in regucalcin transgenic rats was enhanced with aging [10]. Calcium contents in the femoral-diaphyseal and -metaphyseal tissues were reduced in regucalcin transgenic male and female rats with increasing ages (5, 14, 25, or 50 weeks old) [10]. Furthermore, we investigated whether the formation of osteoclastic cells is enhanced in *ex vivo* experiment [10]. Bone marrow cells obtained from wild-type or regucalcin transgenic rats with increasing ages were cultured for 7 days, and the number of TRACP-positive MNCs was markedly increased by culture of bone marrow cells obtained from both groups rats [10]. Of note, the stimulatory effects of PTH (1-34) or 1,25(OH)$_2$D$_3$ on

TRACP-positive MNCs formation in bone marrow cell culture were enhanced in the bone marrow cells from regucalcin transgenic male and female rats with 14 or 25 weeks old [9, 10]. Osteoclastic bone resorption was demonstrated to enhance in regucalcin transgenic rats with aging [10].

Stimulatory effects of regucalcin on osteoclastogenesis were enhanced with aging, suggesting an involvement of regucalcin in the revelation of osteoporosis. Regucalcin is expressed in the bone marrow cells of rats, and it may be released from bone marrow cells. Regucalcin secreted into extracellular parts may exhibit a stimulatory effect on osteoclastogenesis in bone marrow cells. Regucalcin, like a cytokine, directly stimulates osteoclast-like cell formation in bone marrow culture.

As described above, regucalcin plays a physiologic role in the formation of osteoclastic cells and osteoclastic bone resorption *in vitro* and *in vivo*.

REGUCALCIN SUPPRESSES OSTEOBLASTOGENESIS

Osteoblasts are differentiated from bone marrow mesenchymal stem cells, and it promotes bone formation and mineralization to increase bone mass [4]. Bone loss was induced in regucalcin transgenic rats *in vivo* [8], suggesting that regucalcin exhibits stimulatory effects on osteoclastic bone resorption and reveals suppressive effects on osteoblastic bone formation associated with osteoclastic bone resorption. In this section, regucalcin has been sown to exhibit suppressive effects on osteoblastogenesis and mineralization in preosteoblastic cells *in vitro* [17, 18].

Regucalcin Is Expressed in Osteoblastic Cells

Regucalcin has been shown to be expressed in osteoblastic cells. Regucalcin mRNA and its protein were found to express in osteoblastic MC3T3-E1 cells [17, 18]. Regucalcin mRNA expression was regulated by various hormones and cytokines in osteoblastic MC3T3-E1 cells *in vitro*

[17]. Osteoblastic cells in reaching on subconfluence were cultured for 24 or 48 hours in a medium containing either vehicle or various hormones without fetal bovine serum (FBS). Culture with PTH, IGF-I or 17β-estradiol, which exhibits an anabolic effect on osteoblastic cells [16, 17], stimulated regucalcin mRNA expression in osteoblastic cells [17]. Meanwhile, regucalcin mRNA expression in osteoblastic cells was suppressed by the culture with 1, 25(OH)$_2$D$_3$ or TNF-α, stimulators of osteoclastic bone resorption [17].

Tumor necrosis factor-α (TNF- α) may play a role as a negative regulator in regucalcin mRNA expression [17]. Such an effect of TNF-α may be mediated through NF-κB signaling implicated in TNF receptor-associated factor (TRAF6) in osteoblastic cells [17]. Suppressive effects of TNF-α on regucalcin mRNA expression may be mediated via NF-κB signaling in osteoblastic cells. NF-κB binding sites are shown to locate in the promoter region of the regucalcin gene [17, 21]. Furthermore, the stimulatory effects of PTH or IGF-I on regucalcin mRNA expression in osteoblastic cells were suppressed by culture with staurosporine, an inhibitor of protein kinase C, or PD98059, an inhibitor of mitogen-activated protein kinase [17]. These results support the view that regucalcin gene expression is enhanced through intracellular signaling factors involved in various protein kinases in osteoblastic cells.

Regucalcin mRNA expression was also stimulated by 17β-estradiol in osteoblastic cells, and this expression was suppressed by culture with 1,25(OH)$_2$D$_3$ [17]. These results indicate that regucalcin mRNA expression is regulated through various steroid receptors, which their DNA binding sites may be existed in the reguclacin gene promoter region in osteoblastic cells.

Thus, regucalcin mRNA expression was demonstrated to regulate by various hormones and cytokines, which participate in the regulation of osteoblastic bone formation. Regucalcin may play a crucial role in the regulation of cellular functions mediated by various hormones and cytokines in osteoblastic cells.

Regucalcin Suppresses Osteoblastogenesis and Mineralization *In Vitro*

There is growing evidence that regucalcin regulates the functions of osteoblastic cells. Regucalcin has been demonstrated to exhibit suppressive effects on osteoblastogenesis and mineralization in preosteoblastic MC3T3-E1 cells *in vitro* [18]. Of note, regucalcin was shown to decrease the activities of alkaline phosphatase and Ca^{2+}/calmodulin-dependent nitric oxide (NO) synthase in osteoblastic cells [22]. Alkaline phosphatase and NO synthase activities were suppressed by the addition of exogenous regucalcin (10^{-10} - 10^{-8} M) into the enzyme reaction mixture containing the lysate of osteoblastic cells [22]. Alkaline phosphatase is an important enzyme that participates in osteoblastic mineralization [23]. Regucalcin may reveal suppressive effects on osteoblastic mineralization. NO synthase activity was increased by the addition of anti-regucalcin monoclonal antibody (5 or 10 ng/ml) in the enzyme reaction mixture containing the cell lysate in either presence or absence of Ca^{2+}/calmodulin [24]. Regucalcin may play a role as suppressor in the regulation of NO synthase in osteoblastic cells. NO synthase is known to generate NO in various types of cells, and this molecule plays a crucial role in the regulation of cell function [24]. Thus, regucalcin, which is expressed in osteoblastic cells, may play a role in the regulation of osteoblastic cell functions.

Regucalcin did not exhibit significant effects on cell proliferation and apoptosis in osteoblastic cells cultured with short term in a medium containing regucalcin (10^{-10} - 10^{-8} M) without FBS [18]. Targeted disruption of transcription factors runt-related transcription factor 2 (Runx2) results in a complete lack of bone formation owing to maturational arrest of osteoblasts [26]. Runx2 mRNA expression suppressed in osteoblastic cells cultured with the addition of exogenous regucalcin (10^{-10} - 10^{-8} M), although the expressions of α1 (I) collagen and glyceroaldehyde-3-phosphate dehydrogenase (G3PDH) mRNAs were not altered [18]. Suppression of Runx2 mRNA expression caused by extracellular regucalcin may lead to the deterioration of cell differentiation and mineralization in osteoblastic cells.

Culture with regucalcin addition has been shown to suppress IGF-I mRNA expression and stimulate TGF-β1 mRNA expression in osteoblastic MC3T3-E1 cells [18]. IGF-I stimulates bone formation in osteoblastic lineage cells [18]. Suppression of IGF-I may partly lead to the deterioration of osteoblastic bone formation. Meanwhile, TGF-β1 is a potent regulator in the differentiation of osteoblastic cells [27]. Regucalcin-induced enhancement of TGF-β1 mRNA expression may link to the suppression of differentiation of osteoblastic cells. Moreover, the activity of alkaline phosphatase, which participates in bone mineralization [23], was markedly reduced in osteoblastic cells cultured by the addition of regucalcin (10^{-10} - 10^{-8} M) [18]. This reduction may lead to decline of osteoblastic mineralization. Molecular mechanistically, the regulatory effects of extracellular regucalcin on Runx2, IGF-I and TGF-β mRNA expressions and alkaline phosphatase activity may be mediated via intracellular signaling factors, which are generated by the stimulation of extracellular regucalcin in osteoblastic cells.

Prolonged culture with exogenous regucalcin leaded to a remarkable suppression of mineralization in osteoblastic cells [18]. Osteoblastic cells in reaching on subconfluence were cultured with the addition of regucalcin (10^{-10} or 10^{-9} M) for 3, 9, or 18 days, and the results with Alizarin red staining showed that this mineralization was markedly depressed by culture with regucalcin for 3, 9, or 18 days [18]. This result further supports the concept that extracellular regucalcin reveals suppressive effects on cell differentiation and mineralization in osteoblastic cells [18].

Regucalcin, which is expressed in osteoblastic cells, may be released to extracellular fluids. A hypothesis is proposed that the released regucalcin binds to the plasma membranes of osteoblastic cells, and that generates new signaling factors in the cells to regulate various gene expressions and enzyme activities, leading to suppression of mineralization in osteoblastic cells.

Signaling Mechanism of Regucalcin Action in Osteoblastic Cells

The process of intracellular signaling implicated in exhibiting the suppressive effect of extracellular regucalcin on the mineralization in osteoblastic cells *in vitro* has been investigated [15, 18, 22]. Iodinated regucalcin has been shown to bind to the plasma membranes isolated from rat liver *in vitro* [28]. Extracellular regucalcin may bind to the plasma membranes of oteoblastic cells. Specific binding sites for regucalcin may be located on the plasma membranes of osteoblastic cells, and the bounded regucalcin may evoke transmitance of signal(s) into the cell and/or it may be internalized in the cells [22]. Pharmacologic tools were used to determine whether the action of extracellular regucalcin is mediated through intracellular signaling factors in osteoblastic cells [22]. Osteoblastic MC3T3-E1 cells were cultured in the presence of exogenous regucalcin (10^{-10} or 10^{-9} M) without FBS [22]. Cell number was not altered by the addition of regucalcin [22]. Interestingly, culture with exogenous regucalcin caused reductions of protein and DNA contents in osteoblastic cells [22]. Regucalcin-induced decrease in protein contents in osteoblastic cells was clearly blocked by the addition of various kinase inhibitors, including protein kinase C, Ca^{2+}/calmodulin-dependent protein kinase, phosphatidylinositide 3 (PI3) kinase and mitogen-activated protein kinase [22]. These results may support the view that the exhibition of the suppressive effects of extracellular regucalcin on protein content in osteoblastic cells is mediated through various protein kinases, which are implicated in intracellular signaling process. Meanwhile, the decrease in DNA content in osteoblastic cells cultured with extracellular regucalcin was not modulated in the presence of above kinase inhibitors [22]. The effect of regucalcin in decreasing cellular DNA content may be mediated via other signaling mechanisms differed from the action of regucalcin on cellular protein content. Thus, extracellular regucalcin may stimulate the process of signaling implicated in the pathways of various protein kinases in osteoblastic cells.

Moreover, the effects of extracellular regucalcin in decreasing protein and DNA contents in osteoblastic cells were found to be not revealed in the presence of cycloheximide, an inhibitor of protein synthesis, or 5, 6-dichloro-1-β-D-ribofuranosylbenzimidazole (DRB), an inhibitor of transcription activity [22]. These inhibitors exhibited the decrease in protein and DNA contents in osteoblastic cells [22]. Culture with extracellular regucalcin suppressed the mRNA expressions of Runx2, a transcription factor [26, 29], and alkaline phosphatase, a key enzyme of mineralization [30], in osteoblastic cells [22]. These results suggest that extracellular regucalcin-linked intracellular signaling factors lead to the regulation of gene expression in osteoblastic cells [22]. Thus, the extracellular regucalcin-generated intracellular signaling molecules may be transmitted to transcription process in the nucleus of osteoblastic cells.

Of note, the effects of extracellular regucalcin in decreasing protein content in osteoblastic cells were also exhibited in the presence of IGF-I that enhances cell proliferation [22, 31]. Meanwhile, the suppressive effects of extracellular regucalcin on DNA content in osteoblastic cells were not revealed by the addition of IGF-I [22]. Extracellular regucalcin was suggested to partly stimulate other signaling systems but not IGF-I-linked to signaling process in osteoblastic cells. Extracellular regucalcin enhanced TNF-α-induced activation of NF-κB promotor activity in osteoblastic cells, which suppress osteoblastogenesis and mineralization [15]. The signaling process of extracellular regucalcin is linked to activation of NF-κB signaling in osteoblastic cells.

Regucalcin plays a multifunctional role as a suppressor of cell signaling linked to the regulation of protein, DNA and RNA syntheses in various types of cells [13, 14]. Further investigation is needed to identify a specific receptor for regucalcin and its related signaling factors, which mediate the action of extracellular regucalcin in osteoblastic cells. Moreover, it is speculated that extracellular regucalcin is internalized into osteoblastic cells, and intracellular regucalcin suppresses protein and DNA synthesis linked to nuclear function in the cells.

REGUCALCIN STIMULATES ADIPOGENESIS ASSOCIATED WITH OSTEOBLASTOGENESIS IN BONE MARROW CELLS

Bone marrow mesenchymal stem cells (MSCs) are multipotent cells, which give rise to adipocytes and ostoblasts among other cell lineages [25, 32]. In bone marrow, the differentiation of MSCs into adipocytes or osteoblasts is competitively balanced [25, 32]. This is based on the cross talk between manifold signaling pathways including those derived from bone morphogenic proteins, wingless type MMTV integration site (Wnt) proteins, hedgehogs, delta/jagged proteins, fibroblastic growth factors, IGFs, and transcriptional regulators of adipocyte and osteoblast differentiation including peroxisome proliferators-activated receptor-gamma (PPAR gamma) and Runx2 [33, 34].

Regucalcin transgenic rats are shown to display pronounced bone loss and hyperlipidemia [8, 9]. In this context, regucalcin may induce hyperlipidemia *in vivo* by suppressing osteoblast differentiation and stimulating adipogenesis from bone marrow mesenchymal stem cells. Of notw, extracellular regucalcin was demonstrated to reveal suppressive effects on osteoblastogenesis and stimulatory effects on adipogenesis with differentiation in mouse bone marrow culture *in vitro* [35]. Moreover, extracellular regucalcin was found to enhance adipogenesis stimulated by insulin, which is involved in the extracellular signal-related kinase (ERK) pathway in 3T3-L1 adipocytes *in vitro*. Extracellular regucalcin exhibited the suppressive effects on differentiation and mineralization in osteoblastic MC3T3-E1 cells, and it suppressed the expression of Runx2 and alkaline phosphatase mRNAs in the cells *in vitro* [18]. These observations support the concept that extracellular regucalcin suppresses differentiation from bone marrow MSCs to osteoblastic cells, and that it depresses osteoblastic differentiation and mineralization.

Adipocytes are differentiated from bone marrow MSCs [25, 32]. Extracellular regucalcin stimulated differentiation from MSCs to adipocytes. Moreover, extracellular regucalcin stimulated adipogenesis in pre-adipocytes. Importantly, extracellular regucalcin stimulated adipogenesis in mouse 3T3-L1 adipocytes *in vitro* [35]. However, extracellular regucalcin did not cause lipid accumulation in pre-adipocytes and mature adipocytes and not enhance differentiation to mature adipocytes [35]. Thus, extracellular regucalcin did not exhibit a direct stimulatory effect on adipogenesis in 3T3-L1 adipocytes *in vitro*.

Insulin has been shown to potentially stimulate adipogenesis and lipid accumulation in mature adipocytes [36]. Extracellular regucalcin was found to enhance insulin-induced stimulation in adipogenesis and lipid accumulation in mature adipocytes *in vitro*, although it did not reveal a direct stimulatory effect on lipid accumulation in the cells [35]. Extracellular regucalcin may play a role as a modulator in the regulation of adipogenesis stimulated by insulin. Interestingly, the stimulatory effects of extracellular regucalcin on insulin-stimulated adipogenesis were perfectly suppressed by culture with PD98059, an inhibitor of ERK signaling [35], which suppresses the effect of insulin on adipogenesis [34]. Extracellular regucalcin may play a regulatory role in the revelation of the effect of insulin mediated through activation of ERK-related pathway in adipocytes. In addition, PD98059 was shown to suppress the effect of extracellular regucalcin in osteoblastic MC3T3-E1 cells *in vitro* [35].

As described above, extracellular regucalcin binds to the plasma membranes of MSCs and adipocytes, and it enhances sensitivity of insulin binding to the insulin-receptors, although regucalcin may not directly bind to insulin receptors. Moreover, intracellular signaling pathways, which are related to insulin action on adipogenesis, may be enhanced through cross talk with signaling factors that are generated by the binding of regucalcin on adipocyte. Regucalcin-related signaling factor(s) is assumed to modulate pathway of insulin-related ERK signaling. Further mechanism by which extracellular regucalcin regulates ostobalstogenesis and adipogenesis in MSCs remains to be elucidated.

Conclusively, the role of regucalcin in the regulation of bone homeostasis in bone marrow environment was hypothesized in Figure 1. Extracellular regucalcin suppresses osteoblastogenesis and stimulates adipogenesis and osteoclastogenesis in bone marrow. Regucalcin may play a pivotal role in the regulation of bone homeostasis implicated in bone remodeling.

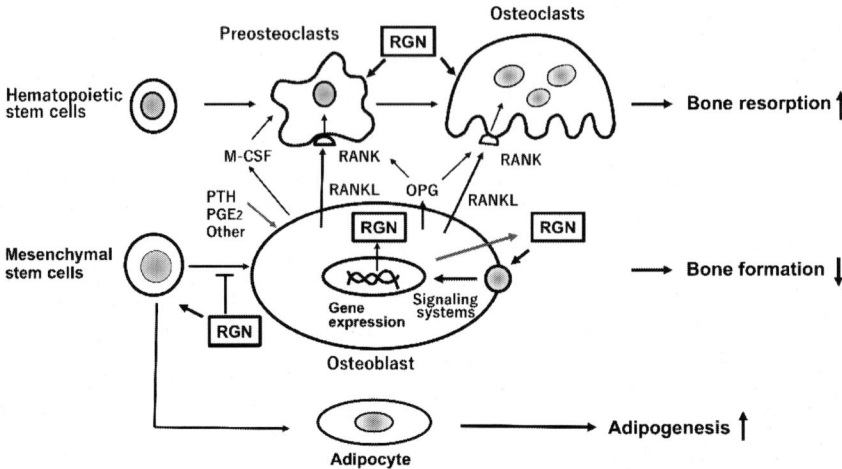

Figure 1. Cellular mechanism of regucalcin in the regulation of bone homeostasis. Extracellular regucalcin, which is secreted from bone marrow cells, may bind to preosteoclasts and stimulate differentiation to osteoclasts. Stimulatory effects of regucalcin on osteoclastogenesis are not implicated in the expression of RNKL or the OPG in osteoblastic cells. Extracellular regucalcin suppresses osteoblastogenesis and mineralization in osteoblasts differentiated from mesenchymal stem cells. Moreover, extracellular regucalcin stimulates adipogenesis in mesenchymal stem cells. The action of extracellular regucalcin is transmitted via signaling system that is linked to various protein kinases. Intracellular regucalcin may regulate signaling factors-related enzyme activation and gene expression in the nucleus of osteoblastic cells. Abbreviations; RGN: regucalcin, RNKL: receptor activator of nuclear factor kappa-B ligand, OPG: osteoprotegerin.

INVOLVEMENT OF REGUCALCIN IN OSTEOPOROSIS

Regucalcin plays a crucial role in the regulation of bone homeostasis. Notably, overexpressed regucalcin was found to induce osteoporosis by using regucalcin transgenic rats *in vivo* [8, 9]. Overexpression of regucalcin was shown to enhance regucalcin mRNA expression and its protein levels in the femoral-diaphyseal and -metaphyseal tissues of rats [8, 9]. These levels were increased in regucalcin transgenic rats [8, 9]. Regucalcin transgenic rats displayed morphologic change in the femoral tissues by using a peripheral quantitative computed tomography (pQCT) [8]. Overexpressed regucalcin evoked a remarkable decrease in mineral content, mineral density and polar strength strain index in the femoral tissues [8]. Thus, regucalcin plays a crucial role in the regulation of bone homeostasis. Meanwhile, the deficiency of endogenous regucalcin has been shown to induce osteomalacia in the regucalcin gene knockout mice *in vivo* [39].

Overexpressed regucalcin leaded to bone metabolic disorder in exhibiting the reduction of alkaline phosphatase activity, a marker enzyme of osteoblastic mineralization [22], in the femoral-diaphyseal and -metaphyseal tissues, indicating a suppression of osteoblastic bone formation [8]. Moreover, the decrease in DNA content in the femoral-diaphyseal and -metaphyseal tissues was displayed in regucalcin transgenic rats [8]. DNA content in the bone tissues may be an index of bone growth and the number of bone cells in femoral tissues. Overexpression of regucalcin leads to suppressions of bone formation and bone growth associated with reduced bone cells (including osteoblastic cells and chondrocytes) in the femoral-metaphyseal tissues of rats. Thus, the findings with biochemical markers and bone morphologic index supported the view that overexpression of regucalcin induces bone loss (osteoporosis). Interestingly, regucalcin transgenic female rats were found to induce a remarkable decrease in bone mass as compared with that of the male rats [8]. Ovarian hormone deficiency with aging is known to lead to osteoporosis with reduction of bone mass [38]. Although the alteration of estrogen expression was not determined in regucalcin transgenic rats, overexpressed regucalcin was demonstrated to enhance bone loss with aging.

Bone loss was occurred in regucalcin transgenic rats with 5 weeks old [8, 10]. The reduction of bone calcium content was caused in the femoral-diaphyseal and -metaphyseal tissues of regucalcin transgenic rats with aging (5 and 36 weeks old). Thus, bone loss was observed in regucalcin transgenic rats with weanling and aging. This result supported the view that bone loss in regucalcin transgenic rats was not restored by bone modeling with aging.

Moreover, a significant elevation of serum inorganic phosphorus concentration, but not calcium, was observed in regucalcin transgenic male and female rats of 5 weeks old [8]. Interestingly, serum inorganic phosphorus, calcium, triglyceride, high-density lipoprotein (HDL)-cholesterol and albumin concentrations were found to significantly increase in regucalcin transgenic female rats with aging (36 weeks old) [8, 10]. A significant alteration of serum lipid components was not observed in regucalcin transgenic rats of 5 weeks old [8, 10]. However, serum triglyceride and HDL-cholesterol concentrations were significantly elevated in the transgenic rats with increasing age [8, 10]. This result suggests that aging induces developed disorder of lipid metabolism, which is associated with bone loss, in regucalcin transgenic rats. Bone marrow mesenchymal stem cells are differentiated to osteoblastic cells and adipocytes [25, 32]. Overexpression of regucalcin may stimulate adipogenesis and induce hyperlipidemia, since regucalcin was shown to stimulate adipogenesis in bone marrow culture *in vitro* [34].

As described above, regucalcin transgenic rats were found to display osteoporosis. This animal model may be useful in the investigation of novel pathogenic mechanism in the progression of osteoporosis. Moreover, the mechanism of cellular events by which overexpression of regucalcin induces bone loss, may provide novel information regarding a mechanistic characterization in the regulation of bone homeostasis.

Meanwhile, regucalcin knockout mice displayed symptoms of scurvy including spontaneous bone fractures, and this was considered to induce with a failure of collagen synthesis owing to vitamin C deficiency [39]. Deficiency of vitamin C may influence on the balance between osteoblasts and osteoclasts in regucalcin knockout mice [40]. Interestingly, the femurs of regucalcin knockout mice displayed decreased bone area and reduced number of osteoblasts as compared with those of wild-type mice and regucalcin knockout mice with vitamin C supplementation [40]. Also, bone alkaline phosphatase expression was decreased in regucalcin knockout mice [40]. In addition, expression of RANKL, which stimulates osteoclastogenesis, was showed to enhance in regucalcin knockout mice [40]. Observations, that regucalcin deficiency induces reduced osteoblastic function and enhanced osteocloastogenesis, suggest that the reduced regucalcin indirectly influences on regulatory systems linked to bone homeostasis [39, 40].

CONCLUSION

The regucalcin gene is shown to express in bone marrow cells and osteoblastic cells [8, 9]. The production of regucalcin in bone marrow including mesenchymal stem cells and osteoblastic cells may be regulated through various hormones and cytokines that regulate bone homeostasis. Extracellular regucalcin, which is produced by bone marrow cells, may impact on mesenchymal stem cells, preosteoclastic cells, osteoblastic cells and adipocytes in bone marrow. Extracellular regucalcin suppresses osteoblastogenesis and stimulates osteoclastogenesis and adipogenesis. Regucalcin may play a crucial role as a cytokine to regulate the differentiation of marrow cells in bone marrow environments [7].

Intracellular regucalcin plays a pivotal role as a suppressor in manifold signaling systems in the regulation of cell function in various types of cells [13, 14]. Intracellular regucalcin expressed in bone marrow cells may play a role in the regulation of cell functions implicated in bone homeostasis. Overexpressed regucalcin leaded to bone loss (osteoporosis) in regucalcin transgenic rats *in vivo* [8-10]. Extracellular regucalcin was found to stimulate osteocalstic bone resorption and to suppress osteoblastogenesis and mineralization *in vitro*, thereby inducing osteoporosis. This bone loss was remarkable in female rats as compared with that of male rats. Osteoporosis is frequently induced in postmenopausal women with estrogen deficiency.

Regucalcin, which plays a pivotal role in the regulation of bone homeostasis, may play a pathophysiologic role in the development of osteoporosis with aging.

ACKNOWLEDGMENT

This study was partly supported by the Foundation for Biomedical Research on Regucalcin (Shizuoka, Japan).

REFERENCES

[1] Parfitt MA (1990) "Bone-forming cells in clinical conditions". In: Hall BK, editor. *Bone Volume 1. The Osteoblast and Osteocyte.* Boca Raton, FL: Telford Press and CRC Press, p. 351-429.
[2] Baron R, Vignery A, Horowitz M (1984) Lymphocytes, macrophages and the regulation of bone remodeling. *Bone Miner Res* 2: 175-243.
[3] Canalis E, McCarthy TI, Centrella M (1988) Growth factors and the regulation of bone remodeling. *J Clinic Invest* 81: 277-281.
[4] Raggatt LJ, Partridge NC (2010) Cellular and molecular mechanisms of bone remodeling. *J Biol Chem* 285: 25103-25108.
[5] Zaidi M, Blair HC, Moonga BS, Abe E, Christopher LH (2003) Osteoclastgenesis, bone resorption, and osteoclast-based therapeutics. *J Bone Miner Res* 18: 599-609.
[6] Chambers TJ, Fuller K (2011) How are osteoclasts induced to resorb bone? *Ann NY Acad Sci* 1240: 1-6.
[7] Yamaguchi M (2014) The role of regucalcin in bone homeostasis: involvement as a novel cytokine. *Integ Biol* 6: 258-266.
[8] Yamaguchi M, Misawa H, Uchiyama S, Morooka Y, Tsurusaki Y (2002). Role of endogenous regucalcin in bone metabolism: Bone loss is induced in regucalcin transgenic rats. *Int J Mol Med* 10: 377-383.
[9] Yamaguchi M, Sawada N, Uchiyama S, Misawa H, Ma ZJ (2004) Expression of regucalcin in rat bone marrow cells: Involvement of osteoclastic bone resorption in regucalcin transgenic rats. *Int J Mol Med* 13: 437-443.

[10] Uchiyama S, Yamaguchi M (2004) Bone loss in regucalcin transgenic rats: Enhancement of osteoclastic cell formation from bone marrow of rats with increasing age. *Int J Mol Med* 14: 451-455.
[11] Yamaguchi M, Uchiyama M (2005) Regucalcin stimulates osteoclast-like cell formation in mouse marrow cultures. *J Cell Biochem* 94: 794-803.
[12] Zou W, Hakim I, Tschoep K, Endres S, Bar-Shavit Z (2001) Tumor necrosis factor-α mediates RANK ligand stimulation of osteoclast-differentiation by an autocrine mechanism. *J Cell Biochem* 83: 70-83.
[13] Yamaguchi M (2000) Role of regucalcin in calcium signaling. *Life Sci* 66: 1769-1780.
[14] Yamaguchi M (2005) Role of regucalcin in maintaining cell homeostasis and function. *Int J Mol Med* 15: 372-389.
[15] Yamaguchi M, Weitzmann MN, Murata T (2012) Exogenous regucalcin stimulates steoclastogenesis and suppresses osteoblastogenesis through NF-κB activation. *Mol Cell Biochem* 359: 193-203.
[16] Asagiri M, Takayanagi H (2007) The molecular understanding of osteoclast differentiation. *Bone* 40: 251-264.
[17] Yamaguchi M, Otomo Y, Uchiyama S, Nakagawa T (2008) Hormonal regulation of regucalcin mRNA expression in osteoblastic MC3T3-E1 cells. *Int J Mol Med* 21: 771-775.
[18] Yamaguchi M, Kobayashi M, Uchiyama S (2005) Suppressive effect of regucalcin on cell differentiation and mineralization in osteoblastic MC3T3-E1 cells. *J Cell Biochem* 96: 543-554.
[19] Jilka RL (2007) Molecular and cellular mechanism of the anabolic effect of intermittent PTH. *Bone* 40: 1434-1446.
[20] Juttner KV, Perry MJ (2007) High-dose estrogen-induced osteogenesis is decreased in aged $RUNX2^{+/-}$ mice. *Bone* 41: 25-32.
[21] Yamaguchi M (2011) The transcriptional regulation of regucalcin gene expression. *Mol Cell Biochem* 346: 147-171.

[22] Otomo Y, Yamaguchi M (2006) Regulatory effect of exogenous regucalcin on cell function in osteoblastic MC3T3-E1 cells: Involvement of intracellular signaling factor. *Int J Mol Med* 18: 321-327.

[23] Majeska RJ, Wuthier RE (1975) Studies on matrix vesicles Isolated from chick epiphyseal cartilage. Association of pyrophosphatase and ATPase activities with alkaline phosphatase. *Biochim Biophys Acta* 391: 51-60.

[24] Lowenstein CJ, Dinerman JL, Snyder SH (1994) Nitric oxide: A physiologic messenger. *Annal Inter Med* 120: 227-237.

[25] Minguel MJ, Erices A, Conget P (2001) Mesenchymal stem cells. *Exp Bioll Med* 226: 507-520.

[26] Komori T, Yagi H, Nomura S, Yamaguchi A, Sasaki K, Deguchi K, Shimizu Y, Bronson RT, Gao YH, Inada M, Sato M, Okamoto R, Kitamura Y, Yoshiki S, Kishimoto T (1997) Targeted disruption of Cbfa1 results in a complete lack of bone formation owing to maturational arrest of osteoblasts. *Cell* 89: 755-764.

[27] Lee MH, Kim YJ, Kim HJ, Park HD, Kang AR, Kyung HM, Sung J H, Wozney JM, Kim HJ, Ryoo HM (2003) BMP-2-induced Runx2 expression is mediated by Dlx5, and TGF-beta 1 opposes the BMP-2-induced osteoblast differentiation by suppression of Dlx5 expression. *J Biol Chem* 278: 34387-34394.

[28] Yamaguchi M, Mori S, Kato S (1988) Calcium-binding protein regucalcin is an activator of (Ca^{2+}-Mg^{2+})-adenosine triphosphatase in the plasma membranes of rat liver. *Chem Pharm Bull (Tokyo)* 36: 3532-3539.

[29] Stock M, Schafer H, Fliegauf M, Otto F (2004) Identification of novel target of the bone-specific transcription factor Runx2. *J Bone Miner Res* 19: 959-972.

[30] Yohay DA, Zhang J, Thrailkill KM, Arthur JA, Quarles DL (1994) Role of serum in the developmental expression of alkaline phosphatase in MC3T3-E1 osteoblasts. *J Cell Physiol* 158: 467-465.

[31] Centrella M, McCarthy TL, Canalis E (1990) Receptors for insulin-like growth factors-I and -II in osteoblast-enriched cultures from fetal rat bone. *Endocrinology* 126: 39-44.

[32] Muruganandan S, Roman AA, Sinal CJ (2009) Adipocyte differentiation of bone marrow-derived mesenchymal stem cells: cross talk with the osteoblastogenesis program. *Cell Mol Life Sci* 66: 236-253.

[33] Zhao L, Hantash BK (2011) TGF-β1 regulates differentiation of bone marrow mesenchymal stem cells. *Vitamin Horm* 87: 127-141.

[34] Laudes MA (2011) Role of WNT signaling in the determination of human mesenchymal stem cells into preadipocytes. *J Mol Endocrinol* 46: R65-R72.

[35] Yamaguchi M, Weitzmann MN, Baile CA, Murata T (2012) Exogenous regucalcin suppresses osteoblastogenesis and stimulates adipogenesis in mouse bone marrow culture. *Integr Biol* 4: 1215-1222.

[36] Hemati NN, Ross SE, Erickson RL, Croblewski CE, MacDougald O A (1997) Signaling pathways through which insulin regulates CCAAT/enhancee binding protein alpha (C/EBPalpha) phosphorylation and gene expression in 3T3-L1 adipocytes. Correlation with GLUT4 gene expression. *J Biol Chem* 272: 25913-25919.

[37] Liu Y, Yang Y, Ye YC, Shi QF, Chai K, Tashiro S, Onodera S, Ikejima T (2012) Activation of ERK-p53 and ERK-mediated phosphorylation of Bcl-2 are involved in autophagic cell death induced by the c-Met inhibitor SU11274 in human lung cancer A549 cells. *J Phramacol Sci* 118: 423-432.

[38] Weitzmann MN, Pacifici R (2006) Estrogen deficiency and bone loss: an inflammatory tale. *J Clin Invest* 116: 1186-1194.

[39] Kondo Y, Inai Y, Sato Y, Handa S, Kubo S, Shimokado K, Goto S, Nishikimi M, Maruyama N, Ishigami A (2006) Senescence marker protein 30 functions as gluconolactonase in L-ascorbic acid biosynthesis, and its knockout mice are prone to scurvy. *Proc Nat Acad Sci USA* 103: 5723-5728.

[40] Park JK, Lee EM, Kim AY, Lee EJ, Min CW, Kang KK, Lee MM, Jeong KS (2012) Vitamin C deficiency accelerates bone loss inducing an increase in PPAR-γ expression in SMP30 knockout mice. *Int J Exp Pathol* 93: 332-340.

In: Regucalcin
Editor: Masayoshi Yamaguchi

ISBN: 978-1-53616-172-4
© 2019 Nova Science Publishers, Inc.

Chapter 5

INVOLVEMENT OF REGUCALCIN IN DIABETES AND HYPERLIPEDEMIA

Masayoshi Yamaguchi[*]
Department of Pathology and Laboratory Medicine,
David Geffen School of Medicine, University of California,
Los Angeles (UCLA), Los Angeles, CA, US

ABSTRACT

Regucalcin, which plays a multifunctional role in the regulation in various types and tissues, may play a pathophysiological role in metabolic disorder. The expression of regucalcin is stimulated through action of insulin in liver cells *in vitro* and *in vivo,* and it is diminished in the liver of rats with type I diabetes after streptozotocin administration *in vivo*. Overexpressed regucalcin stimulates glucose utilization and lipid production in liver cells with glucose supplementation *in vitro*. Regucalcin exhibits insulin resistance in liver cells. Deficiency of regucalcin induces an impairment of glucose tolerance and lipid accumulation in the liver of mice *in vivo*. Overexpressed regucalcin stimulates glucose utilization and lipid production in modeled rat hepatoma H4-II-E cells. Overexpressed regucalcin enhances glucose

---

[*] Corresponding Author's E-mail: yamamasa11555@yahoo.co.jp.

transporter 2 mRNA expression to stimulate glucose utilization, and it depresses the gene expression of insulin receptor or phosphoinositide 3-kinase involved in insulin signaling, which is enhanced by insulin and/or glucose supplementation, leading to insulin resistance in liver cells. Moreover, overexpressed regucalcin decreases in triglyceride, total cholesterol and glycogen contents in the liver of rats *in vivo*, inducing a hyperlipidemia. Leptin and adiponectin mRNA expressions in the liver tissues are suppressed in regucalcin transgenic rats. The decrease in hepatic regucalcin is associated with the development and progression of nonalcoholic fatty liver disease and fibrosis in human patients. Regucalcin may be a key molecule implicated in diabetes and lipid metabolic disorder.

Keywords: regucalcin, glucose metabolism, lipid metabolism, diabetes, hyperlipidemia

INTRODUCTION

Regucalcin was discovered in 1978 as a unique calcium-binding protein that does not contain EF-hand motif of calcium-binding domain [1]. The regucalcin gene (gene symbol; *rgn*) is localized on the X chromosome, and it is identified in over 15 species consisting of regucalcin family [2, 3]. Regucalcin has been shown to play a multifunctional role in cell regulation; maintaining of intracellular calcium homeostasis and suppressing of signal transduction, translational protein synthesis, nuclear deoxyribonucleic acid (DNA) and ribonucleic acid (RNA) synthesis, proliferation, and apoptosis in various types of cells [4-7].

There is growing evidence that the change in regucalcin gene expression in the fat bodies of insect are linked with photoperiodic regulation of cold tolerance in insect (Drosophila montana) [7], suggesting that regucalcin plays a crucial role in energy metabolism under cold environmental condition. Regucalcin may play a pivotal role in the regulation of energy metabolism in invertebrate and vertebrate specifies conserved with evolutional development. In fact, overexpression of regucalcin was demonstrated to induce hyperlipidemia related to obesity and diabetes, as introduced in this chapter.

Calcium (Ca^{2+}) plays a pivotal role in the regulation of various intracellular metabolic pathways including glucose metabolism and diabetes [9]. Regucalcin has been shown to reverse the activation of various enzymes by Ca^{2+} in glycogenesis and glycolysis in the liver [10-13]. Regucalcin is focused to play a role in glucose metabolism in the liver. There is accumulating evidence that regucalcin may be a key molecule in metabolic disorder including glucose and lipid metabolism *in vivo* [14, 15]. Moreover, regucalcin is demonstrated to stimulate adipogenesis and differentiation to adipocytes in bone marrow cells [16], suggesting a physiologic role of regucalcin in the regulation of lipid metabolism. Notably, hyperlipidemia is caused in regucalcin transgenic rats *in vivo* [17].

This chapter discusses current topics regarding the role of regucalcin in the regulation of glucose and lipid metabolisms and involvement of diabetes and hyperlipidemia.

INVOLVEMENT OF REGUCALCIN IN INSULIN RESISTANCE

Insulin plays a pivotal role in the regulation of glucose and lipid metabolisms. Regucalcin mRNA expression and its protein level are expressed in modeled human liver cancer HepG2 cells [18]. Insulin has been shown to increase regucalcin mRNA expression and its protein levels in HepG2 cells *in vitro* [18]. Moreover, insulin increased regucalcin gene expression in the liver of rats fasted for overnight *in vivo* [19]. Fasting decreased markedly regucalcin mRNA expression in the liver [19]. Refeeding produced a remarkable increase in regucalcin mRNA expression in the liver [19]. Liver regucalcin levels were not changed by fasting, and were markedly increased by refeeding [19]. Moreover, regucalcin mRNA expression in the liver was increased by the oral administration of glucose (2 g/kg body weight) to fasted rats [19], indicating an involvement of insulin secreted from pancreatic cells after glucose administration. Furthermore, hepatic regucalcin mRNA expression was raised by a single subcutaneous administration of insulin (10 and 100 U/kg body weight) to fasted rats *in vivo* [19]. These results suggest that insulin plays a role in stimulation of the expression of regucalcin *in vitro* and *in vivo*, suggesting that regucalcin involves in the regulation of insulin action.

The resistance of insulin action may lead to the condition of diabetes. Notably, regucalcin is shown to involve in insulin resistance [20]. Insulin resistance is generated in modeled rat hepatoma H4-II-E cells cultured with tumor necrosis factor-alpha (TNF-α) and insulin [20]. *In vitro* model nicely mimicked insulin resistance observed in human type 2 diabetic mellitus. In this model study, H4-II-E cells were cultured with either insulin, TNF-α, or TNF-α plus insulin [20]. Interestingly, regucalcin was identified as a protein, which is involved in insulin resistance by using proteome analysis of H4-II-E cells exposed to insulin and TNF-α [20]. There were identified other 13 proteins including eukaryotic translation initiation factor-3, subunit 2, regulator of G-protein signaling-5, superoxide dismutase, protein disulfide isomerase A6, proteasome subunit-alpha type 3, disulfide isomerase A6, cell-division protein kinase-4, kinogen heavy chain, carbonic anhydrase-7, E 3 ubiquitin protein ligase, URE-B1, Rab GDP dissociation inhibitor-beta, Rab GDP dissociation inhibitor-beta2, and MAWDBP [20]. These proteins are known as regulators of translation, protein degradation, cellular Ca^{2+} signaling, G-proteins, and free-radical production [20].

Regucalcin has been shown to regulate the action of insulin on glucose and lipid metabolism. Regucalcin was found to regulate glucose and lipid metabolism in liver cells by using modeled rat hepatoma H4-II-E cells *in vitro* [21]. Overexpressed regucalcin stimulated the production of triglyceride and free fatty acid in H4-II-E cells cultured with or without the supplementation of glucose in the absence of insulin [21]. This result suggests that regucalcin stimulates lipid production linked to glucose metabolism in the cells *in vitro*. In addition, insulin enhanced an increase in the consumption of glucose and productions of triglyceride and free fatty acid in wild-type cells cultured with glucose supplementation [21]. This effect of insulin was suppressed by overexpression of regucalcin [21]. Regucalcin may suppress the action of insulin on glucose and lipid metabolisms in liver cells, suggesting that the action of regucalcin leads to insulin resistance in liver cells.

Whether regucalcin reveals regulatory effects on the expression of various enzymes related to glucose and lipid metabolism in the liver has been investigated. The expressions of acetyl-CoA carboxylase, HMG-CoA reductase, glucokinase, pyruvate kinase, and glyceroaldehyde-3-phosphate dehydrogenase mRNAs in wild-type cells was not significantly altered by culture with or without glucose supplementation in the presence of insulin [21]. The expressions of these gene were not altered by the overexpression of regucalcin [21]. Thus, regucalcin did not have effects on the gene expression of various enzymes, which are related to glucose and lipid metabolisms, in H4-II-E cells. However, regucalcin has been shown to exhibit the regulatory effect on the activity of various enzymes linked to glucose and lipid metabolisms in H4-II-E cells [10-13].

Overexpression of regucalcin has been shown to elevate glucose transporter 2 (GLUT 2) mRNA expression in modeled H4-II-E cells [21]. This elevation was not modulated in regucalcin-overexpressing cells cultured in the presence of insulin with or without glucose supplementation [21]. Thus, GLUT 2 mRNA expression is stimulated by overexpressed regucalcin, and the enhanced expression may lead to the increase in glucose utilization in H4-II-E cells.

Furthermore, we investigated the effect of overexpressed regucalcin on the gene expression of insulin signaling-related proteins by using regucalcin-overexpressing H4-II-E cells *in vitro* [22]. The expression of rat insulin receptor (Insr) or phosphatidylinositol 3-kinase (PI3K) mRNAs was enhanced in the cells cultured with glucose supplementation in the presence of insulin [22]. This enhancement was suppressed by overexpression of regucalcin [22]. These suppressive effects were not seen in the cells cultured in the absence of insulin [22]. Thus, overexpression of regucalcin was found to suppress the gene expression of insulin signaling-related proteins including Insr and PI3K. These results suggest that regucalcin plays a crucial role in inducing insulin resistance in liver cells.

Involvement of Regucalcin in Diabetes

Regucalcin plays a pathophysiologic role in diabetes. The content of regucalcin in the liver was shown to decrease in type-I diabetic state induced by the administration of streptozotocin (STZ) [23]. STZ (60 mg/kg body weight) was subcutaneously administered in rats, and 1 or 3 weeks later they were sacrificed by bleeding. Regucalcin mRNA expression in the liver was not altered in diabetic state [23]. However, liver regucalcin levels were markedly reduced by STZ treatment [23]. Serum regucalcin concentration was not altered by STZ treatment [23]. These results suggest that the reduced hepatic regucalcin is not resulted from a decrease in liver regucalcin mRNA expression and an increase in the release of regucalin into the serum. Hepatic regucalcin was found to decrease in STZ-diabetic rats.

Insulin has been shown to stimulate both regucalcin mRNA and its protein levels in the liver cells [18, 19]. The secretion of insulin in pancreatic cells was impaired in STZ diabetic state, and the impaired production of insulin may lead to reduction in regucalcin levels in the liver with diabetic state. The activity of serum transaminases was elevated in STZ-diabetic state, suggesting that liver injury is induced in diabetic state [23]. The decrease in hepatic regucalcin with diabetic state may lead to the disorder of liver metabolism.

Deficiency of regucalcin induces an impairment of glucose tolerance in regucalcin knockout mice treated with L-ascorbic acid, because the deficiency of regucalcin causes a decline in serum vitamin C [24, 25]. Blood glucose levels were elevated, and serum insulin level was reduced in regucalcin knockout mice intraperitoneally administered with glucose as compared with those of wild-type mice [23]. In this glucose tolerance test, the regucalcin knockout mice showed a greater glucose-lowering effect [23], indicating that the knockout mice reveal enhanced insulin sensitivity regardless of vitamin C status [23]. Moreover, high-fat diet feeding severely was shown to worsen glucose tolerance in both wild type and regucalcin knockout mice [23]. In incubation study of islets *ex vivo*, the secretion of insulin in response to glucose or potassium chloride was reduced in islets from regucalcin knockout mice [23]. Islet adenosine triphosphate content in regucalcin knockout mice was similar to that in wild-type mice [23]. These results suggest that regucalcin deficiency impairs insulin secretion [23]. Reduced regucalcin may lead to the worsening of glucose tolerance by decreasing insulin secretion [23]. Regucalcin may play a stimulatory role in insulin secretion in pancreatic cells. Insulin has been shown to stimulate the expression of regucalcin mRNA in modeled hepatoma H4-II-E cells [18, 19].

The role of vitamin C on insulin secretion from pancreatic β-cells has been investigated using regucalcin knockout mice that induces vitamin C-deficient [26]. Regucalcin knockout mice impaired glucose tolerance with the decrease in blood insulin levels as compared with that of wild-type mice [26]. In contrast, vitamin C-sufficient regucalcin knockout mice showed a higher blood glucose and lower insulin levels [26]. This result shows that the deficiency of regucalcin reveals reduced insulin sensitivity regardless of vitamin C status [26].

Deficiency of regucalcin may impair insulin secretion, and this may lead to the development of diabetes.

INVOLVEMENT OF REGUCALCIN IN LIPID METABOLISM AND HYPERLIPIDEMIA

There is growing evidence that regucalcin plays a crucial role in the regulation of lipid metabolism. Culture with exogenous regucalcin has been shown to stimulate adipogenesis in bone marrow cells *ex vivo* [16], suggesting an involvement of regucalcin in lipid metabolism. Furthermore, it has been investigated an involvement of regucalcin in lipid metabolism by using the regucalcin gene engineered animal models *in vivo*. Hepatocytes from regucalcin knockout mice, but not the wild-type mice at 12 months of age, were shown to contain many lipid droplets, abnormally enlarged mitochondria with indistinct cristae, and enlarged lysosomes filled with electron-dense bodies in the electron microscope [27].

Biochemical analysis of neutral lipids, total triglyceride and cholesterol in the liver from regucalcin knockout mice showed higher levels as compared with those from age-matched wild-type mice [27]. Deficiency of regucalcin was found to induce accumulation of neutral lipids and phospholipids including phosphatidylethanolamine, cardiolipin, phosphatidylcholine, phosphatidylserine and sphingomyelin in the liver in mice [27]. These results support the view that regucalcin participates in the regulation of lipid metabolism in the liver of mice.

Notably, regucalcin transgenic rats, which overexpresses endogenous regucalcin, has been shown to display bone loss associated with elevation of serum triglyceride and high-density lipoprotein (HDL)-cholesterol concentrations at the age of 36 weeks *in vivo* [28, 29]. Serum free fatty acid, triglyceride, cholesterol or HDL-cholesterol concentration was markedly elevated in regucalcin transgenic male and female rats at 14 - 50 weeks of age [28]. Serum calcium concentration was raised in regucalcin transgenic male and female rats at 50 weeks of age [28]. Serum albumin concentration was significantly increased in regucalcin transgenic female rats at 25-50 weeks of age [28]. Serum zinc, glucose or urea nitrogen concentrations were not significantly changed in transgenic male and female rats [28]. These results demonstrate that hyperlipidemia is uniquely induced in regucalcin transgenic rats with increasing age *in vivo*. Thus, overexpressed regucalcin leads to hyperlipidemia in rats *in vivo*.

Overexpressed regucalcin was found to induce the change in lipid components in the adipose and liver tissues [29]. Regucalcin is expressed in the adipose tissues of normal rats [29]. Regucalcin expression in the adipose tissues was not increased in the transgenic rats [29]. Overexpressed regucalcin stimulated the release of lipid components from the adipose tissue into the serum [29]. Triglyceride content in the adipose tissues was increased in 50-week-old regucalcin transgenic rats, while the free fatty acid content in the adipose tissues was not altered [29]. Elevated lipids in the adipose tissues of regucalcin transgenic rats may be released into the serum.

The expression of regucalcin in the liver tissues was increased in regucalcin transgenic rats [29]. Hepatic triglyceride, total cholesterol, free fatty acid or glycogen content was reduced in regucalcin transgenic rats [29]. Regucalcin has been shown to reveal suppressive effects on the activations of glycogen particulate phosphorylase *a* [12], cytoplasm pyruvate kinase [11] and fructose 1, 6-diphosphatase [10] by Ca^{2+} and calmodulin in rat liver. Overexpressed regucalcin may lead to depression of glycogen synthesis and promotion of glycogenolysis in the liver of rats. Furthermore, overexpressed regucalcin may stimulate lipid synthesis in the liver tissues of transgenic rats *in vivo*.

Overexpression of regucalcin has been shown to regulate glycolytic metabolism in rat prostate [30]. Glucose consumption was decreased in the prostate of regucalcin transgenic rats as compared with that of wild-type rats, and accompanied by reduced expression of GLUT 3 and glycolytic enzyme phosphofructokinase [30]. Overexpressed regucalcin reduced lactate levels, which resulted from the diminished expression and/or activity of lactate dehydrogenase, in the prostates of rats [30]. Moreover, overexpression of regucalcin was shown to decrease the expression of monocarboxylate transporter 4 involved in the export of lactate to the extracellular space [30]. These results suggest that regucalcin suppresses glycolytic metabolism in rat prostate, and that the loss of regucalcin is suggested to predispose to a hyperglycolytic profile [30].

Leptin and adiponectin are adipocytokines implicated in lipid metabolism [31, 32]. Leptin mRNA expression in the adipose or liver tissues was reduced in regucalcin transgenic rats at 50-week-old [29]. Adiponectin mRNA expression was not significantly altered in the adipose tissues of regucalcin transgenic rats at 50-week-old, but interestingly it was reduced in the liver tissues [29]. These decreases may lead to hyperlipidemia in regucalcin transgenic rats.

As described above, our results of *in vitro* [21, 22] and *in vivo* [28, 29] experiments provide a possible mechanism by which regucalcin induces hyperlipidemia, as summarized in Figure 1.

Hyperlipidemia has been known to display in lipoprotein lipase-deficient mice [33], low-density lipoprotein (LDL) receptor-deficient mice [34], apolipoprotein C3-KO mice [35], apolipoprotein C1 TG mice [36], very LDL lipoprotein receptor KO mice [37], cholesterol 7 alpha-hydroxylase-deficient mice [38], apoE-deficient mice [39], and liver myr-Akt overexpressing mice [40]. These animal models for hyperlipidemia are implicated in molecules that regulate lipid metabolism. Regucalcin transgenic rats may be an interesting as an animal model for hyperlipidemia. Regucalcin may be a key molecule that regulates lipid metabolism.

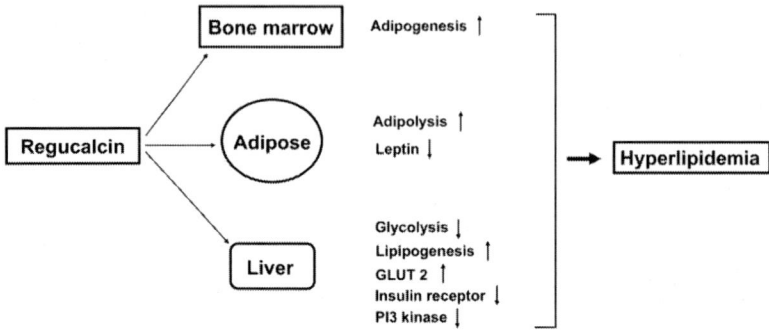

Figure 1. Overexpressed regucalcin induces disorder of lipid metabolism in the adipose and liver tissues. Triglyceride content in the adipose tissues is elevated in regucalcin-overexpressing transgenic rats, and triglyceride, total cholesterol and free fatty acid contents in the liver tissues are reduced in transgenic rats. Hyperlipidemia, which is displayed in regucalcin transgenic rats, is proposed to be induced by the elevation of lipid components in the liver and the depression of leptin or adiponectin gene expressions in the adipose or liver.

Involvement of Regucalcin in Nonalchoholic Fatty Liver Disease

The expression of regucalcin in the liver has been shown to depress with condition of liver disease. Liver regucalcin mRNA expression and its protein levels were reduced by the administration of carbon tetrachloride [41], galactosamine [42], and ethanol [43], which induces liver metabolic disorder in rats *in vivo*. Liver regucalcin was released into the serum with liver damage [41, 42]. Regucalcin was found to have a potential sensitivity as a marker of chronic liver disease in human subjects [44]. Liver disease, which is implicated to lipid metabolic disorder, is further developed through the reduction of regucalcin expression in the liver.

The involvement of regucalcin in liver fibrosis has been investigated using regucalcin knockout mice. Carbon tetrachloride administration-induced liver fibrosis and the nuclear translocation of phosphorylated Smad2/3, which is signaling factors of transforming growth factor-β, was found to depress in the liver of regucalcin knockout mice as compared with that of wild-type mice [45]. Regucalcin was not expressed in hepatic stellate cells (HSCs) of mice [46]. Peroxisome proliferator-activated receptor-gamma (PPAR-γ) was shown to upregulate in the liver of regucalcin knockout mice [46]. Numerous HSCs were hypertrophic and contained abundant microvesicular lipid droplets in the liver cytoplasm of aged regucalcin knockout mice [46]. The expression of PPAR-γ, which is a protein related to lipid metabolism and HSC quiescence, was elevated in hypertrophic HSCs of regucalcin knockout mice [46]. Reduced regucalcin may lead to development to liver fibrosis.

Regucalcin is a molecule implicated in insulin resistance [14, 15, 20]. Insulin resistance in the liver was associated with the pathogenesis of nonalcoholic fatty liver disease (NAFLD). Change in hepatic regucalcin levels may be associated with the development and progression of NAFLD [47]. Patients with NAFLD showed a significant lower level of liver regucalcin [47]. Liver regucalcin levels were suppressed in a fibrosis stage-dependent manner and were negatively implicated with the assessment of insulin resistance, the net electronegative charge modified-low-density lipoprotein (LDL), and type IV collagen 7S [47]. The immune-staining intensity levels of 4-hydroxynonenal in the liver were inversely associated with hepatic regucalcin levels [46]. Both serum large very LDL and very small LDL levels were found to increase in patients with NAFLD [47]. Whether or not regucalcin decreased in the liver of human patients is a result or a cause of cirrhosis, remains to be elucidated [47]. However, suppressed regucalcin may lead to NAFLD and cirrhosis in the liver of human.

CONCLUSION

Regucalcin plays a crucial role in the regulation of glucose and lipid metabolism [15]. Moreover, regucalcin possess a pathophysiological role in diabetes and lipid metabolic disorder. Importantly, regucalcin expression in the liver is enhanced by insulin, and regucalcin is identified to be a molecule implicated in insulin resistance in liver cells. Moreover, the deficiency and overexpression of endogenous regucalcin has been shown to induce lipid metabolic disorder and diabetes. Regucalcin may be a key molecule to understand pathophysiology of lipid metabolic disorder.

Human studies of regucalcin involved in diabetes and hyperlipidemia will be important to elucidate implication with clinical aspects. Regucalcin may be a target molecule for therapy of diabetes and lipid metabolic disorder in patients.

ACKNOWLEDGMENT

This study was partly supported by the Foundation for Biomedical Research on Regucalcin (Shizuoka, Japan).

REFERENCES

[1] Yamaguchi M, Yamamoto T (1978) Purification of calcium binding substance from soluble fraction of normal rat liver. *Chem Pharm Bull (Tokyo)* 26:1915-1918.

[2] Shimokawa N, Yamaguchi M (1993) Molecular cloning and sequencing of the cDNA coding for a calcium-binding protein regucalcin from rat liver. *FEBS Lett* 327: 251-255.

[3] Shimokawa N, Matsuda Y, Yamaguchi M (1995) Genomic cloning and chromosomal assignment of rat regucalcin gene. *Mol Cell Biochem* 151:157-163.

[4] Yamaguchi M (2000) Role of regucalcin in calcium signaling. *Life Sci* 66: 1769-1780.
[5] Yamaguchi M (2005) Role of regucalcin in maintaining cell homeostasis and function. *Int J Mol Med* 15: 372-389.
[6] Yamaguchi M (2011) Regucalcin and cell regulation: role as a suppressor in cell signaling. *Mol Cell Biochem* 353:101-137.
[7] Yamaguchi M (2011) The transcriptional regulation of regucalcin gene expression. *Mol Cell Biochem* 346: 147-171.
[8] Vesala L, Salminen TS, Kankare M, Hoikkala A (2012) Photoperiodic regulation of cold tolerance and expression levels of regucalcin gene in Drosophila Montana. *J Insect Physiol* 58:704-709.
[9] Kraus-Friedman N, Li F (1980) The role of intracellular Ca^{2+} in the regulation of glucogenesis. *Metabolism* 48: 389-403.
[10] Yamaguchi M, Yoshida H (1985) Regulatory effect of calcium-binding protein isolated from rat liver cytosol on activation of fructose 1, 6-diphosphatase by Ca^{2+}-calmodulin. *Chem Pharm Bull (Tokyo)* 33: 4489-4493.
[11] Yamaguchi M, Shibano H (1987) Calcium-binding protein isolated from rat liver cytosol reverses activation of pyruvate kinase by Ca^{2+}. *Chem Pharm Bull (Tokyo)* 35: 2025-2029.
[12] Yamaguchi M, Shibano H (1987) Effect of calcium-binding protein on the activation of phosphorylase *a* in rat hepatic particulate glycogen by Ca^{2+}. *Chem Pharm Bull (Tokyo)* 35: 2581-2584.
[13] Yamaguchi M, Mori S, Suketa Y (1989) Effects of Ca^{2+} and V^{5+} on glucose-6-phosphatase activity in rat liver microsomes: The Ca^{2+} effect is reversed by regucalcin. *Chem Pharml Bull (Tokyo)* 37:388-390.
[14] Yamaguchi M (2010) Regucalcin and metabolic disorder: osteoporosis and hyperlipidemia are induced in regucalcin transgenic rats. *Mol Cell Biochem* 327: 53-63.
[15] Yamaguchi M, Murata T (2013) Involvement of regucalcin in lipid metabolism and diabetes. *Metabolism* 62:1045-1051.
[16] Yamaguchi M, Weitzmann MN, Baile CA, Murata T (2012) Exogenous regucalc suppresses osteoblastogenesis and stimulates

adipogenesis in mouse bone marrow culture. *Integr Biol* 4:1215-1222.

[17] Yamaguchi M, Misawa H, Uchiyama S, Morooka Y, Tsurusaki Y (2002) Role of endogenous regucalcin in bone metabolism: Bone loss is induced in regucalcin transgenic rats. *Int J Mol Med* 10: 377-383.

[18] Murata T, Shinya N, Yamaguchi M (1997) Expression of calcium-binding protein regucalcin mRNA in the cloned human in hepatoma cells (Hep G2): Stimulation by insulin. *Mol Cell Biochem* 175: 163-168.

[19] Yamaguchi M, Oishi K, Isogai M (1995) Expression of hepatic calcium-binding protein regucalcin mRNA is elevated by refeeding of fasted rats: Involvement of glucose, insulin and calcium as stimulating factor. *Mol Cell Biochem* 142:35-41.

[20] Solomon S S, Buss N, Shull J, Monnier S, Majumdar G, Wu J, Gerling IC (2005) Proteome of H-411E (liver) cells exposed to insulin and tumor necrosis factor-alpha: analysis of proteins involved in insulin resistance. *J Lab Clin Med* 145:275-283.

[21] Nakashima C, Yamaguchi M (2006) Overexpression of regucalcin enhances glucose utilization and lipid production in cloned rat hepatoma H4-II-E cells: Involvement of insulin resistance. *J Cell Biochem* 99:1582-1592.

[22] Nakashima C, Yamaguchi M (2007) Overexpression of regucalcin suppresses gene expression of insulin signaling-related proteins in cloned rat hepatoma H4-II-E cells: involvement of insulin resistance. *Int J Mol Med* 20:709-716.

[23] Isogai M, Kurota H, Yamaguchi M (1997) Hepatic calcium-binding protein regucalcin concentration is decreased by streptozotocin-diabetic state and ethanol ingestion in rats. *Mol Cell Biochem* 168:67-72.

[24] Hasegawa G, Yamasaki M, Kadono M, Tanaka M, Asano M, Senmaru T, Kondo Y, Fukui M, Obayashi H, Maruyama N, Nakamura N, Ishigami A (2010) Senescence marker protein-30/gluconolactonase deletion worsens glucose tolerance through impairment of acute insulin secretion. *Endocrinology* 151:529-536.

[25] Hasegawa G (2010) Decreased senescence marker protein-30 could be a factor that contributes to the worsening of glucose tolerance in normal aging. *Islets* 2:258-260.

[26] Senmaru T, Yamasaki M, Okada H, Asano A, Fukui M, Nakamura N, Obayashi H, Kondo Y, Maruyama N, Ishigami A, Hasegawa G (2012) Pancreatic insulin release in vitamin C-deficient senescence marker protein-30/gluconolactonase knockout mice. *J Clin Biochem Nutr* 50:114-118.

[27] Ishigami A Kondo Y, Nanba R, Ohsawa T, Handa S, Kubo S, Akita M, Naoki M (2004) SMP30 deficiency in mice causes an accumulation of neutral lipids and phospholipids in the liver and shortens the life span. *Biochim Biophys Res Commun* 315:575-580.

[28] Yamaguchi M, Igarashi A, Uchiyama S, Sawada N (2004) Hyperlipidemia is induced in regucalcin transgenic rats with increasing age. *Int J Mol Med* 14: 647-651.

[29] Yamaguchi M, Nakagawa T (2007) Change in lipid components in the adipose and liver tissues of regucalcin transgenic rats with increasing age: Suppression of leptin and adiponectin gene expression. *Int J Mol Med* 20: 323-328.

[30] Vaz CV, Marques R, Cardoso HJ, Maia CJ, Socorro S (2016) Suppressed glycolytic metabolism in the prostate of transgenic rats overexpressing calcium-binding protein regucalcin underpins reduced cell proliferation. *Transgenic Res* 25:139-148.

[31] Havel PJ (2004) Update on adipocyte hormones: regulation of energy balance and carbohydrate/lipid metabolism. *Diabetes* 53: 5143-5151.

[32] Roni T, Lupattelli G, Mannarino E (2006) The endocrine function of adipose tissue: an uptake. *Clin Endocrinol* 64: 355-365.

[33] Weinstock, PH, Bisgaier CL, Aalto-Setala K, Radner H, Ramarkrishnan R, Levak-Frank S, Essenbury AD, Zechner R, Breslow JL (1995) Severe hypertriglyceridemia, reduced high density lipoprotein, and neonatal death in lipoprotein lipase in knockout mice: Mild hypertriglyceridemia with impaired very low density lipoprotein clearance in heterozygotes. *J Clin Invest* 96: 2555-2568.

[34] Lichtman AH, Clinton SK, Liyama K, Connelly PW, Libby P, Cybulsky MI (1999) Hyperlipidemia and atherosclerotic lesion development in LDL receptor-deficient mice fed defined semipurifdied diets with and without cholate. *Arter Thromb Vascul Biol* 19: 1938-1944.

[35] Jong MC, Havekes LM (2000) Insights into apolipoprotein C metabolism from transgenic and gene-targeted mice. *Int J Tissue React* 22: 59-66.

[36] Koopmans SJ, Jong MC, Que I, Dahlmans VEH, Pijl H, Radder JK, Frolich M, Havekes LM (2001) Hyperlipidemia is associated with increased insulin-mediated glucose metabolism, reduced fatty acid metabolism and normal blood pressure in transgenic mice overexpressing human apolipoprotein C1. *Diabetologia* 44: 437-443.

[37] Yagyu H, Lutz EP, Kako Y, Marks S, Hu Y, Choi SY, Bensadoun A, Goldberg IJ (2002) Very low density lipoprotein (VLDL) receptor-deficient mice have reduced lipoprotein lipase activity mass with VLDL receptor deficiency. *J Biol Chem* 277: 10037-10043.

[38] Chen JY, Levy-Wilson B, Goodart S, Cooper AD (2002) Mice expressing the human CYP7A1 gene in the mouse CYP7A1 knock-out background lock induction of CYP7A1 expression by cholesterol feeding and have increased hypercholesterolemia when fed a high fat diet. *J Biol Chemi* 277: 42588-42595.

[39] Fazio S, Linton MF (2001) Mouse models of hyperlipidemia and Atheroscerosis. *Front Biosci* 6: D515-D525.

[40] Ono H, Shimano H, Katagiri H, Yahagi N, Sakoda H, Onishi Y, Anai M, Ogihara T, Fujishiro M, Viana AYI, Fukushima Y, Abe M, Shojima N, Kikuchi M, Yamada N, Oka Y, Asano T (2003) Hepatic Akt activation induces marked hypoglycemia, hepatomegaly, and hypertriglyceridemia with sterol regulatory element binding protein involvement. *Diabetes* 52: 2905-2913.

[41] Isogai M, Shimokawa N, Yamaguchi M (1994) Hepatic calcium-binding protein regucalcin is released into the serum of rats administered orally carbon tetrachloride. *Mol Cell Biochem* 131:174-179.

[42] Isogai M, Oishi K, Yamaguchi M (1994) Serum release of hepatic calcium-binding protein regucalcin by liver injury with galactosamine administration in rats. *Mol Cell Biochem* 136:85-90.

[43] Isogai M, Kurota H, Yamaguchi M (1997) Hepatic calcium-binding protein regucalcin concentration is decreased by streptozotocin diabetic state and ethanol ingestion in rats. *Mol Cell Biochem* 168:67-72.

[44] Yamaguchi M, Isogai M, Shimada N (1997) Potential sensitivity of hepatic specific protein regucalcin as a marker of chronic liver injury. *Mol Cell Biochem* 167: 187-190.

[45] Jeong DH, Hwang M, Park JK, Goo MJ, Hong IIH, Ki MR, Ishigami A, Kim AY, Lee EM, Lee EJ, Jeong KS (2013) Smad3 deficiency ameliorates hepatic fibrogenesis through the expression of senescence marker protein-30, an antioxidant-related protein. *Int J Mol Sci* 14:23700-23710.

[46] Hong IIH, Han JY, Goo MJ, Hwa SY, Ki MR, Park JK, Hong KS, Hwang OK, Kim TH, Yoo SE, Jeong KS (2012) Ascorbic acid deficiency accelerates aging of hepatic stellate cells with up-regulation of PPARγ. *Histol Histopathol* 27:171-179.

[47] Park H, Ishigami A, Shima T, Mizuno M, Maruyama N, Yamaguchi K, Mitsuyoshi H, Minam Mi, Yasui K, Itoh Y, Yoshikawa T, Fukui M, Hasegawa G, Nakamura N, Ohta M, Obayshi H, Okamoto T (2010) Hepatic senescence marker protein-30 is involved in the progression of nonalcoholic fatty liver disease. *J Gastroenterol* 45:426-434.

In: Regucalcin
Editor: Masayoshi Yamaguchi

ISBN: 978-1-53616-172-4
© 2019 Nova Science Publishers, Inc.

Chapter 6

EXTRACELLULAR VESICLES: A CONVERSATION IN THE BODY

Naoomi Tominaga[*]
Department of Biology, Massachusetts Institute of Technology,
Cambridge, MA, US

ABSTRACT

Extracellular vesicles, which are nano mater scale vesicles, are made in cells through some pathways. As one of the important pathway, these are secreted from multivesicular bodies fusion with the plasma membrane. It is contained many functional molecules such as DNAs, mRNAs, microRNAs, and proteins. These molecules are transferred from the donor cells to the other recipient cells to change the environment and/or to send some signals from cell to cell, which seems like a conversation of the cells. The extracellular vesicles are related to the malignancy of cancer, the development of neuron, and the immune system. Regucalcin also found in the extracellular vesicles from urine, which are downregulated in the state of the diabetic nephropathy. It is expected to apply for the early diagnosis of disease. In this way, through about four decades, many researchers had been uncovering the function of the

---

[*] Corresponding Author Email: ntominag@mit.edu.

extracellular vesicle. On the other hand, the extracellular vesicles were named as "Exosome" at first as the definition of "the released vesicles which may serve a physiologic function". However, the extracellular vesicles are named very different terminology through about four decades. In this chapter, it will be an overview the history, terminology, collection methods, contents and function of extracellular vesicles.

Keywords: extracellular vesicles (EVs), exosome, regucalcin

TERMINOLOGY AND HISTORY

Terminology

Extracellular vesicles (EVs) are nanometer vesicles, which secrete from cells (Figure 1). EVs are included many types of vesicles such as exosome, ectosome, microvesicle, oncosome, and the other vesicles. There are two types of EVs based on the origin which is exosomes and microvesicles.

The exosome has used the terminology of the microvesicles about 100nm which secreted from cells through the research of EVs. The exosomes come from the multivesicular bodies origin from the endosome, which is about 50-150 nm [1]. On the other hand, the microvesicle is budded from the plasma membrane. The size has a variety from 50 nm to 500 nm (up to 1 μm).

In the recent year, the terminology is confused because many researchers named vesicles which based on each them; however, these vesicles are collected from the supernatant of cell culture medium or body fluid in most case. The term for EVs has a different claim based on the collection methods. However, it is difficult to distinguish each other. Furthermore, it seems like one of the causes of confusing that the "exosome" has been used as a multienzyme ribonuclease complex. It is not used as a released vesicle [2].

In response to this situation, the International Society for Extracellular Vesicles, established in 2011, recommended using the term "extracellular vesicle(s)" as a hypernym for all types of vesicles present in the extracellular space at 2013 [3].

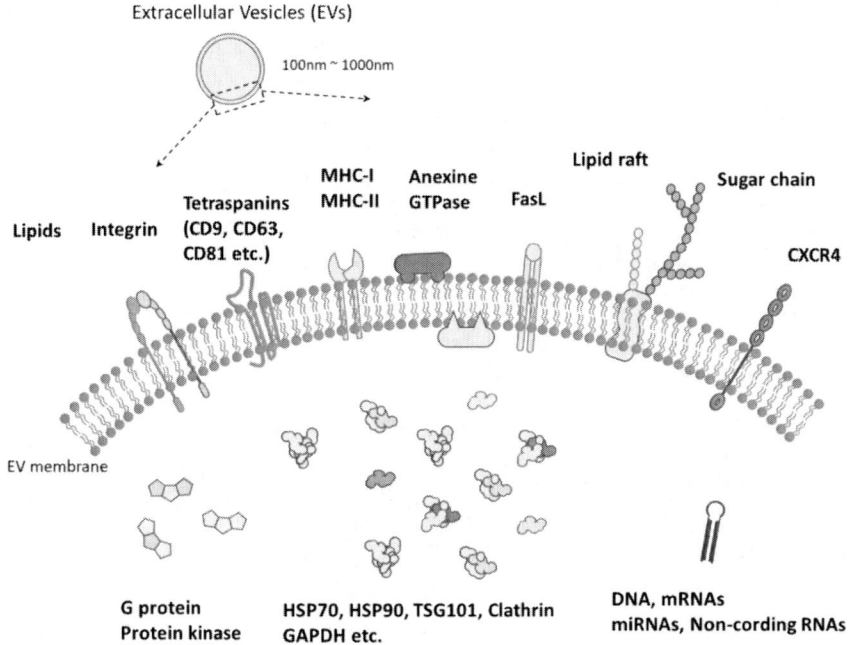

Figure 1. The structure and contents of extracellular vesicles. The extracellular vesicle consists of a lipid bilayer. The extracellular vesicle contains many functional molecules, such as DNAs, mRNAs, miRNAs, G proteins, Heat Shock Proteins. There are many proteins, such as Tetraspanins, Integrins, Major histocompatibility complexes (MHCs), on the lipid bilayer of the extracellular vesicle. The extracellular vesicles are secreted from cells. These extracellular vesicles are uptaken by recipient cells, and then these functional molecules are work in the recipient cells.

History

The research history of extracellular vesicle (include exosome) dates back by about 40 years ago (Figure 2). It has been consistently called exosome, early in history. In 1977, Hans Lutz et al. reported the released vesicles from old sheep erythrocyte which shaped headlike [4]. The same year, the other group report a functional fraction in the supernatant of the prostatic fluid, which seems like a fraction of EVs [5]. Lutz et al. are also reported the mechanism of vesicle release from human red blood cells in 1981 [6, 7]. Until then the existence of protein on released vesicles from erythrocyte was known; however, in 1983, Pan and Johnston characterized the contents of secreted vesicles [8]. They reported that vesicles were secreted from sheep reticulocyte during maturation which contained some lipids and proteins. The secreted vesicles are collected from the supernatant of culture media by centrifuge at 100,000xg after removing cell debris. They reported that this secretion of vesicles is increased time-dependent by detecting with ^{125}I-labeled anti-transferrin receptor in the supernatant. In this article, they use 'released vesicles', 'externalized vesicles,' 'small extracellular vesicle'.

Afterward, two groups reported that the transferrin receptor externalization to multivesicular bodies [9-11]. In 1987, Johnston et al. used "exosome" which is released vesicles from multivesicular bodies during reticulocyte maturation [12]. Actually, it had been proposed that the term "exosome" is referred for the released vesicles which may serve a physiologic function in 1981 by Eberhard G. Trams et al. [13]. After that, the existence of exosomes unveiled in the circulation of several species which include human [14, 15]. In 1991, however, it was concluded that exosomes were the externalization of unnecessary membrane proteins through reticulocyte maturation [16]. It seems like to be one of the reasons why exosomes were considered "garbage bin". Afterward, Rab4 and ARF were suggested to relate to exosome formation [17], and the contents of exosomes such as complement receptor 1 [18], heat shock protein-70 [19], major histocompatibility complex [20], glycosylphosphatidylinositol (GPI)-anchored proteins [21, 22], and Integrin α4β1 [23] were reported. Around 2000, it was reported that some proteins were selectively sorted to vesicles and released into extracellular milieu [21, 22, 24]. In 2001, Wolfers et al. first reported that tumor-derived exosome which contained tumor antigen promoted the anti-tumor effect by dendritic cells [25]. Through 2003 to 2005, it was proposed that exosome related immune response [26-28] which can be useful for immunotherapy [27-32]. In this way, the extracellular vesicle (exosome) research has been studying in the limelight through related to cancer immune response.

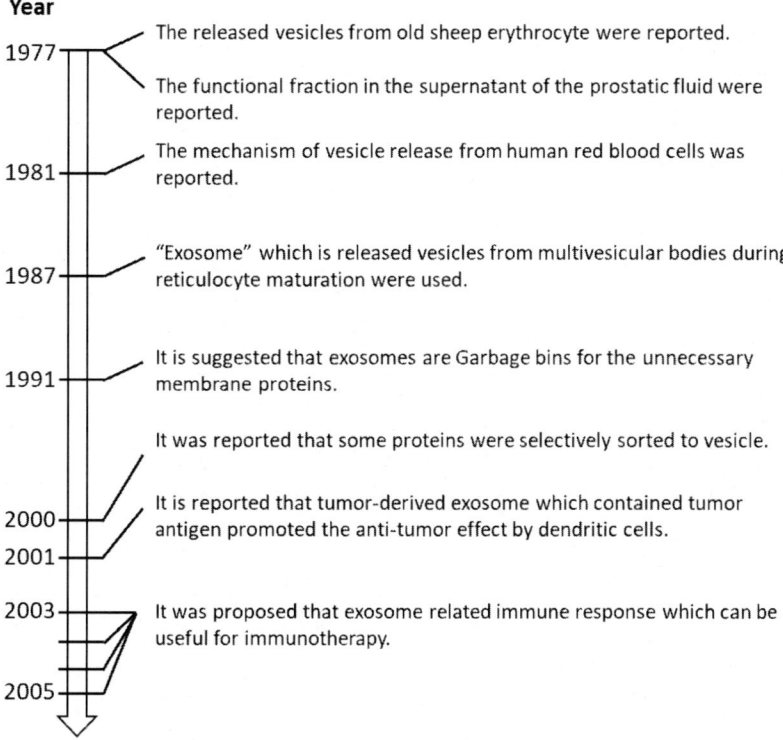

Figure 2. The history of research of extracellular vesicles. The extracellular vesicles were first reported around 1977. After that, the evidence of function for extracellular vesicle has been accumulating. It had revealed that the extracellular vesicles could be useful for cancer immunotherapy during 2003-2005. After that, the extracellular vesicle research has been studying in the limelight through related to cancer immune response. However, the terminology of the extracellular vesicle is confused in the recent year.

ISOLATION OF EXTRACELLULAR VESICLES

There are many types of methods for the isolation of EVs, such as ultracentrifugation [33], density gradient centrifugation [34], filtration [34], immunoaffinity [35]. It is suggested that "Choice of the method(s) should be guided by the required degree of EV purity and concentration [36]." The earliest papers of EVs had used ultracentrifugation method for purification [4-7]. Furthermore, the ultracentrifugation method is the most popular method in recent year for now. It is reported that the EVs have biodistribution for some organs, which is collected with the ultracentrifugation method [37]. Importantly, the biodistribution of EVs is different between the collection methods [37, 38]. These results suggest that the EVs by ultracentrifugation include heterogeneous populations. For the reasons, it is essential which method will be chosen for the EVs collection. Even so, the ultracentrifugation method for collection of EVs would be the first step for the purification. In this section, the method of EVs by ultracentrifugation from the supernatant of cell culture will be introduced [39].

The cells, which utilize on research, will be seed on the culture plate about 80% confluence after the routine cell culture (Figure 3). The culture medium should use the low- or non-fetal bovine serum contained culture medium (e.g., Advanced-DMEM or Advanced-RPMI1640 culture medium which to allow for serum reduction) to avoid contamination of EVs from the fetal bovine serum. The cells, which culture in the low- or non-fetal bovine serum culture medium, will be incubated 48 hours in the routine condition such as in 5% CO_2 at 37°C. This incubation time depends on the state of cells. The culture medium will be centrifuged with low speed because of to remove the cell debris and large particles. The supernatant will be cautiously collected to a new tube which avoids to contamination of precipitated pellets after the centrifugation. This supernatant after low-speed centrifugation will be filtered with 0.22 μm filter, which is an advanced removing for the cell debris and large particles contamination. In the next step, the supernatant will be performed ultracentrifugation at 100,000xg for 70 minutes. After the ultracentrifugation, the supernatant will be discarded, then PBS will be added to the ultracentrifuge tube for washing step. The tubes contain PBS will be performed additional ultracentrifugation at 100,000 x g for 70 minutes. In the case of the density gradient centrifugation, it will be used 1.15-1.19 g/mL sucrose. The EVs should be collected at the bottom of the tubes.

Collection of supernatant contained EVs

☐	Seed on the culture plate about 80% confluency
☐	Incubate overnight at 37'C, 5% CO2
☐	Wash with non-FBS culture medium, next day
☐	Add the right amount of non-FBS culture medium
☐	Incubate 48 hrs at 37'C, 5% CO2
☐	Collect supernatant after 48hrs incubation

Concentration of EVs from the supernatant

☐	Centrifuge of collected supernatant with low-speed
☐	Move supernatant to a new tube, but a precipitate should not contaminate.
☐	Filtration the collected supernatant with 0.22 μm filter
☐	Centrifuge of the filtrate supernatant at 100,000xg for 70 min
☐	Supernatant discard after super-centrifugation; EVs are concentrated at the bottom of the tube.

Figure 3. The protocol of concentration of extracellular vesicles from the culture medium.

FORMATION AND SECRETION

Formation of the Multivesicular Body

EVs are originated two pathways which are (1) secreted from the multivesicular bodies, which is also called multivesicular endosomes, or (2) shed from the plasma membrane [40] (Figure 4A).

Multivesicular bodies are contained EVs which are generated as intraluminal vesicles. The origin of the intraluminal vesicles is generated in early endosome through endosomal maturation from Golgi or is internalized from the plasma membrane to intraluminal vesicles [41]. There are two different pathways in the formation of intraluminal vesicles via endosomal maturation which are the endosomal sorting complex required for transport (ESCRT)-dependent or ESCRT-independent pathway [1].

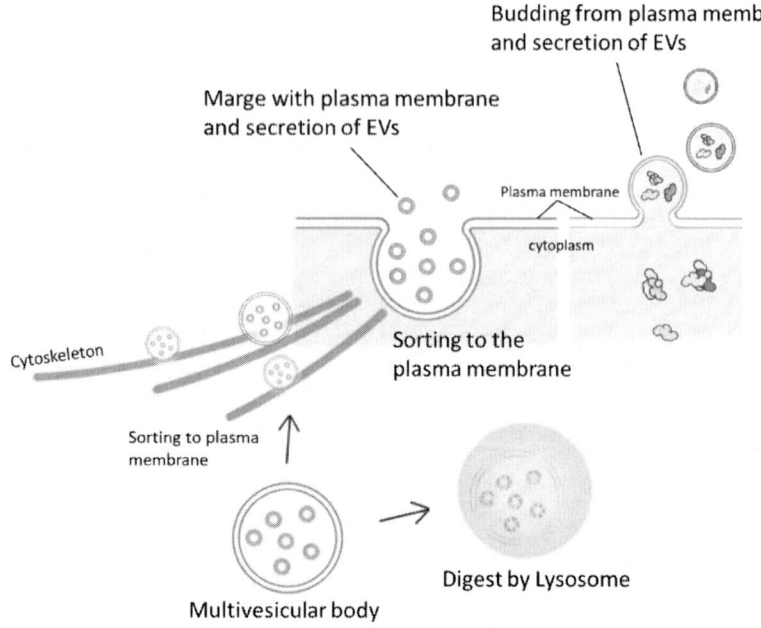

Figure 4A. Secretion pathway of extracellular vesicles. There are two different pathways for the origins of extracellular vesicles. It is the intraluminal vesicles which consist of the multivesicular body. The other is the budding vesicles from the plasma membrane. The multivesicular body sort to the plasma membrane through on the cytoskeleton, these are secreted to extracellular space after fused between multivesicular body and plasma membrane.

In the case of the ESCRT-dependent pathway, the membrane cargos on the lumen of the multivesicular body are assembled by the machinery of ESCRT-0 and Clathrin [42] (Figure 4B). The cytoplasmic sorting occurs in the next step, and then these pre-intraluminal vesicles are organized with ESCRT-I and ALG-2 interacting protein X (ALIX) which is an ESCRT accessory protein [43]. Finally, the pre-intraluminal vesicles are dissociated from the lumen of the multivesicular body by ESCRT-III to become the intraluminal vesicles [43].

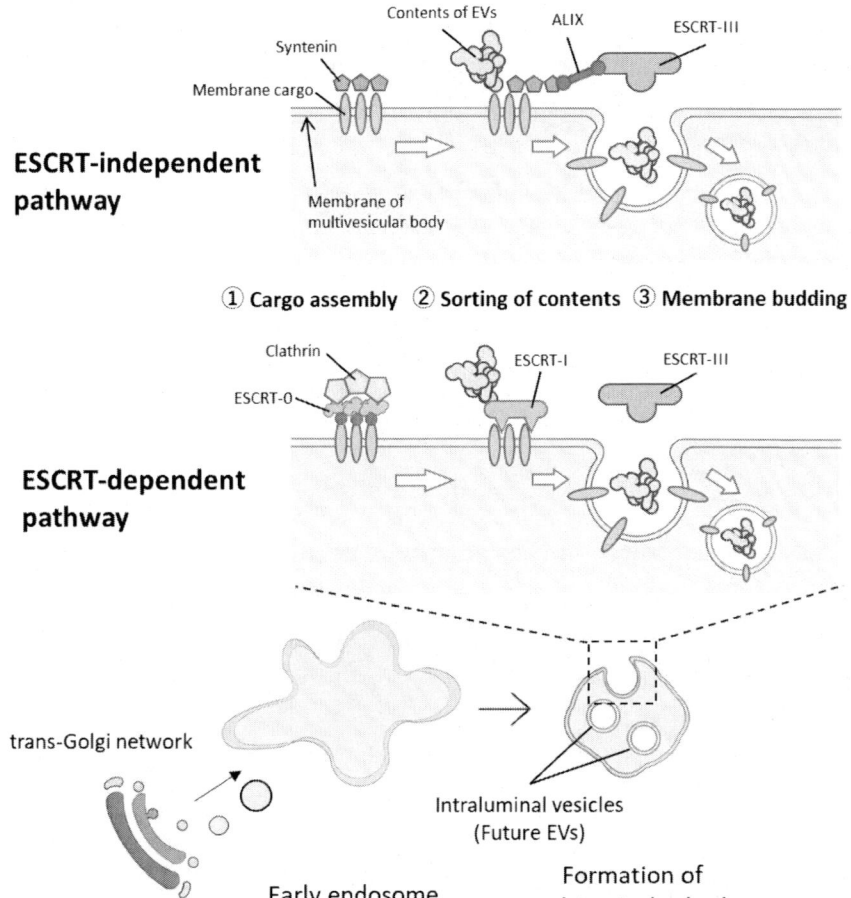

Figure 4B. The formation mechanism of the multivesicular body. It is formed that the multivesicular body from the early endosome through the trans-Golgi network. The intraluminal vesicles, which is a future extracellular vesicle, are formed on the lumen of the early endosome. There are two pathways in the formation of the intraluminal vesicle, which is ESCRT-independent and ESCRT-dependent pathway.

ESCRT machinery is an essential protein complex for the membrane budding from cytoplasm which consisted of ESCRT-0, -I, -II, and –III. ESCRT machinery is found in many organisms which play the formation of the multivesicular body [40], viral budding [44], nuclear pore complex quality control [45], nuclear envelope reassembly [46, 47], nuclear envelope repair after rupture [48], and plasma membrane repair [49].

On the other hand, in case of the ESCRT-independent pathway, the membrane cargos on the lumen of the multivesicular body are assembled by Syntenin [43] and Tetraspanin [50, 51]. The sorting to the pre-vesicles is performed in the assembly step. ALIX and ESCRT-III form the intraluminal vesicles[43]. The ceramide which hydrolyzes from sphingomyelin by neutral type II sphingomyelinase is needed for the formation of EVs in ESCRT-independent pathway [52, 53]. The tetraspanin family, such as CD63, CD81, CD82, and CD9, is working in the sorting of various contents to the EVs [48, 50, 51, 54].

Secretion Mechanism of Extracellular Vesicles

The multivesicular bodies have to fuse with the plasma membrane to release the intracellular vesicles which are the EVs. However, there are multi-steps until the fusion of the multivesicular bodies with the plasma membrane [1]. As mentioned above, the multivesicular bodies are formed with some mechanisms. There are three routes for the multivesicular bodies after biogenesis which are 1) to the Lysosome [55], 2) to the autophagosome [56], and 3) to the plasma membrane [57]. The multivesicular bodies have destinated the degradation when fusing with lysosome or autophagosome. The inhibition of endosomal proton pump V-ATPase read to increase secretion of the EVs by a deterioration of lysosome activity [58-59]. The extracellular vesicle secretion is also increased by prion protein which inhibits the formation of autophagosome [60]. The secretion of EVs is increased by chemical inhibition of autophagy [61].

The intracellular transportation of multivesicular bodies to the plasma membrane is performed by dynein, kinesins, and myosins on the cytoskeleton [62, 63]. The RAB-GTPase such as RAB7, RAB27A, RAB27B, RAB11, and RAB35 are essential for the secretion of intracellular vesicles [43, 64, 65]. RAB27A and B regulate the motility of multivesicular body, and the rearrange of actin to fuse the plasma membrane. Inhibition of RAB27s leads the result of decrease intracellular vesicle secretion.

THE FUNCTION OF EXTRACELLULAR VESICLES

The Contents and Biological Functions of Extracellular Vesicles

The EVs contain many types of molecules such as lipids, proteins, mRNAs, microRNAs, and DNAs [66, 67] (Figure 1). These molecules have a biological function. These EVs are secreted from many types of cells, such as mesenchymal stem cells [68], nerve cells [69], oligodendrocytes [70], Schwann cells [71], immune cells (such as T cells, B cells, macrophages, dendritic cells) [1], oocyte [72, 73], and so on. Moreover, these EVs are related to many types of disease, such as cancer [74, 76], respiratory disease [77], circulatory disease [78], infection disease [79], Alzheimer's disease [80], diabetes [81], kidney disease [82] and so on [37].

As mentioned above, the EVs from immune cells are contributing to an expansion of extracellular vesicle research. For example, the EVs from T cells promote own proliferation by paracrine and promote apoptosis of cancer cells which have the fatty acid synthase ligand (FASL) by the fatty acid synthase (FAS) on the EVs [83]. The EVs from B cells are uptake by macrophages [84]. The function of such macrophages affects the state of EVs in the bloodstream.

Interestingly, the EVs from mesenchymal stem cells, which is an origin from mesoderm, promote a cell proliferation for an organ which injured [68]. In case of a skin burn, these EVs promote cell proliferation of skin cells, which cause of Wnt4 on the EVs [85].

The oligodendrocytes also secrete the EVs. For example, the EVs from oligodendrocytes are uptake by nerve cells selectively [70]. These EVs contribute to the protection of nerve cells.

Furthermore, it was revealed that the EVs from pancreatic β-cells contributed type-II diabetes through the chronic inflammation of the pancreas [81]. In pathophysiological conditions, the pancreatic β-cells secrete EVs contain miRNAs. These EVs from the pancreatic β-cells are uptaken by the neighboring pancreatic β-cells. As a result, the transferred EVs induce apoptosis of the cells. On the other hand, it is reported that the EVs from pancreatic β-cells inhibit apoptosis of the pancreatic β-cells by low-dose inflammatory cytokine [86].

In summary, the EVs from cells are secreted from cells; then these vesicles are uptaken by cells by paracrine and autocrine mechanisms. Some EVs are uptaken by cells which are far away from the donor cells of EVs through the bloodstream. The contents of EVs are work in the recipient cells such as cancer metastasis, immune response, inhibit or promote the cell apoptosis, and both. It is predicted that the EVs useful for drugs, treatments. Furthermore, some EVs are discharged from the kidney through urea. Therefore, it is predicted that EVs will be useful for diagnosis.

The Application of Extracellular Vesicles for the Diagnosis of Kidney Disease

It is possible to use the EVs from urine for the diagnosis. It is known as 'liquid biopsy' which is non-/low-invasive diagnosis method compared with the canonical biopsy [87, 88]. The method of using urine is one of the non-invasive diagnosis methods. The chronic kidney disease is included such as nephritis, diabetic nephropathy, chronic glomerulonephritis, and nephrosclerosis. Focal glomerulosclerosis is a kidney disease which shows a nephrotic syndrome. The EVs from the urine of the patient of focal glomerulosclerosis are increased content of the Wilm's tumor 1 (WT1) protein [89]. In the case of nephritis, it is reported that microRNA-26a is increased [90], and microRNA-29c has decreased the contents of the EVs from urine [91]. In the case of diabetic nephropathy which is the frequent cause of end-stage renal disease, it is reported that WT1, alpha-1-

microglobulin/bikunin precursor (AMBP), histone lysine N-methyltransferase (MLL3), and microRNAs are increased the contents of the EVs from urine [92] [93]. On the other hand, the voltage-dependent anion-selective channel protein 1 (VDAC1), miR-155, and miR-424 have decreased the contents of the EVs from the urine of the patients of diabetic nephropathy [92, 94]. Furthermore, it is reported that different microRNA can be detected between type-1 and type-2 diabetes. MiR-130a, miR-145, miR-155, and miR-424 can be detected in the EVs in urine from the diabetic nephropathy of type-1 diabetes. miR-15b, miR-30, miR-34a, miR-133b, miR-320c, miR-342, and miR-636 can be detected in the EVs in urine from the diabetic nephropathy of type-2 diabetes [95, 96]. Interestingly, regucalcin downregulated in the EVs from the urine of diabetic nephropathy patients compared with healthy donors [97]. The regucalcin was also strongly downregulated in the kidney tissue of diabetic nephropathy patients compared with the healthy donor. The amount of regucalcin in EVs are collated with cells. Regucalcin from EVs in urine is expected for early diagnosis of diabetic kidney disease.

CONCLUSION

The EVs are secreted from many types of cells in the body. Importantly, it is reported that EVs have many biological activities. The regucalcin found in the EVs from the healthy donor's urine; however, it is decreased in the EVs in the urine from the diabetic nephropathy patients. From these results, it is expected that the regucalcin exist in the extracellular vesicle on the bloodstream which secreted from many types of cells in the body. As mentioned above, EVs are incorporated in recipient cells; these some molecules are transferred to the recipient cells from donor cells. Therefore, it is also expected that the amount of regucalcin in the EVs may go up and down in various body conditions, be taken up by cells, and may be functioning in donor cells. Importantly, EVs research still unknown in detail; on the other hand, it is hoped that EVs are useful for the diagnosis. Further research is needed to unveil about the EVs.

REFERENCES

[1] Niel G. van, D'Angelo G, Raposo G (2018) Shedding light on the cell biology of extracellular vesicles. *Nat Rev Mol Cell Biol* 19: 213–228.

[2] Mitchell P, Petfalski E, Shevchenko A, Mann M, Tollervey D (1997) The Exosome: A Conserved Eukaryotic RNA Processing Complex Containing Multiple 3'→5' Exoribonucleases. *Cell.* 91: 457–466.

[3] Gould SJ, Raposo G (2013) As we wait: coping with an imperfect nomenclature for extracellular vesicles. *J Extracell Vesicles* 2:3–5. https://doi.org/10.3402/jev.v2i0.20389.

[4] Lutz HU, Lomant AJ, McMillan P, Wehrli E (1977) Rearrangements of integral membrane components during in vitro aging of sheep erythrocyte membranes. *J Cell Biol* 74: 389–398.

[5] Ronquist G, Hedström M (1977) Restoration of detergent-inactivated adenosine triphosphatase activity of human prostatic fluid with concanavalin A. *Biochim Biophys Acta* 483: 483–486.

[6] Müller H, Schmidt U, Lutz HU (1981) On the mechanism of vesicle release from ATP-depleted human red blood cells. *Biochim Biophys Acta* 649: 462–470.

[7] Zweig SE, Tokuyasu KT, Singer SJ (1981) Member-associated changes during erythropoiesis. On the mechanism of maturation of reticulocytes to erythrocytes. *J Supramol Struct Cell Biochem* 17: 163–181.

[8] Pan BT, Johnstone RM (1983) Fate of the transferrin receptor during maturation of sheep reticulocytes in vitro: selective externalization of the receptor. *Cell* 33: 967–978.

[9] Harding C., Heuser J., Stahl P. (1983) Receptor-mediated endocytosis of transferrin and recycling of the transferrin receptor in rat reticulocytes., *J. Cell Biol.* 97: 329–339.

[10] Harding C, Heuser J, Stahl P (1984) Endocytosis and intracellular processing of transferrin and colloidal gold-transferrin in rat reticulocytes: demonstration of a pathway for receptor shedding. *Eur J Cell Biol* 35: 256–263.

[11] Pan BT, Teng K, Wu C, Adam M, Johnstone RM (1985) Electron microscopic evidence for externalization of the transferrin receptor in vesicular form in sheep reticulocytes. *J Cell Biol* 101: 942–948.

[12] Johnstone RM, Adam M, Hammond JR, Orr L, Turbide C (1987) Vesicle formation during reticulocyte maturation. Association of plasma membrane activities with released vesicles (exosomes). *J Biol Chem* 262: 9412–9420.

[13] Trams EG, Lauter CJ, Salem N, Heine U (1981) Exfoliation of membrane ecto-enzymes in the form of micro-vesicles. *Biochim Biophys Acta* 645: 63–70.

[14] Johnstone RM, Bianchini A, Teng K (1989) Reticulocyte maturation and exosome release: transferrin receptor containing exosomes shows multiple plasma membrane functions. *Blood* 74: 1844–1851.

[15] Johnstone RM, Ahn J (1990) A common mechanism may be involved in the selective loss of plasma membrane functions during reticulocyte maturation., *Biophy Biochim Acta* 49: S70-S75.

[16] Johnstone RM, Mathew A, Mason AB, Teng K (1991) Exosome formation during maturation of mammalian and avian reticulocytes: evidence that exosome release is a major route for externalization of obsolete membrane proteins. *J Cell Physiol* 147: 27–36.

[17] Vidal MJ, Stahl PD (1993) The small GTP-binding proteins Rab4 and ARF are associated with released exosomes during reticulocyte maturation. *Eur J Cell Biol* 60: 261–267.

[18] Pascual M, Lutz HU, Steiger G, Stammler P, Schifferli JA (1993) Release of vesicles enriched in complement receptor 1 from human erythrocytes., *J. Immunol.* 151: 397–404.

[19] Mathew A, Bell A, Johnstone RM (1995) Hsp-70 is closely associated with the transferrin receptor in exosomes from maturing reticulocytes. *Biochem J* 308: Pt 3: 823–830.

[20] Raposo G, Nijman HW, Stoorvogel W, Liejendekker R, Harding CV, Melief CJ, Geuze HJ (1996) B lymphocytes secrete antigen-presenting vesicles. *J Exp Med* 183: 1161–1172.

[21] Rabesandratana BH, Toutant J, Reggio H, Vidal M (1998) Decay-Accelerating Factor (CD55) and Membrane Inhibitor of Reactive

Lysis (CD59) Are Released Within Exosomes During In Vitro Maturation of Reticulocytes. *Blood* 91: 2573–2581.

[22] Escola JM, Kleijmeer MJ, Stoorvogel W, Griffith JM, Yoshie O, Geuze HJ (1998) Selective enrichment of tetraspan proteins on the internal vesicles of multivesicular endosomes and on exosomes secreted by human B-lymphocytes. *J Biol Chem* 273: 20121–20127.

[23] Rieu S, Géminard C, Rabesandratana H, Sainte-Marie J, Vidal M (2000) Exosomes released during reticulocyte maturation bind to fibronectin via integrin alpha4beta1. *Eur J Biochem* 267: 583–590.

[24] Théry C, Regnault A, Garin J, Wolfers J, Zitvogel L, Ricciardi-Castagnoli P, Raposo G, Amigorena S (1999) Molecular characterization of dendritic cell-derived exosomes. Selective accumulation of the heat shock protein hsc73. *J Cell Biol* 147: 599–610.

[25] Wolfers J, Lozier A, Raposo G, Regnault A, Théry C, Masurier C, Flament C, Pouzieux S, Faure F, Tursz T, Angevin E, Amigorena S, Zitvogel L (2001) Tumor-derived exosomes are a source of shared tumor rejection antigens for CTL cross-priming. *Nat Med* 7: 297–303.

[26] Pêche H, Heslan M, Usal C, Amigorena S, Cuturi MC (2003) Presentation of donor major histocompatibility complex antigens by bone marrow dendritic cell-derived exosomes modulates allograft rejection. *Transplantation* 76: 1503–1510.

[27] Hwang I, Shen X, Sprent J (2003) Direct stimulation of naive T cells by membrane vesicles from antigen-presenting cells: distinct roles for CD54 and B7 molecules. *Proc Natl Acad Sci USA* 100: 6670–6675.

[28] Skokos D, Botros HG, Demeure C, Morin J, Peronet R, Birkenmeier G, Boudaly S, Mécheri S (2003) Mast cell-derived exosomes induce phenotypic and functional maturation of dendritic cells and elicit specific immune responses *in vivo*. *J Immunol* 170: 3037–3045.

[29] Altieri SL, Khan ANH, Tomasi TB (1997) Exosomes from plasmacytoma cells as a tumor vaccine. *J Immunother* 27: 282–288

[30] Carpentier AF, Lemonnier F, Adema G, Chaput N, Aubert N, Andre F, Angevin E, Taieb J, Schartz NEC, Tursz T, Adams M, Merad M,

Bonnaventure P, Novault S, Zitvogel L, Escudier B, Bernard J, Ferrantini M (2014) Exosomes as potent cell-free peptide-based vaccine. II. exosomes in CpG adjuvants efficiently prime naive Tc1 lymphocytes leading to tumor rejection. *J Immunol* 172: 2137–2146.

[31] Chaput N, Taïeb J, Schartz NEC, André F, Angevin E, Zitvogel L, Exosome-based immunotherapy. *Cancer Immunol Immunother* (2004) 53: 234–239.

[32] Escudier B, Dorval T, Chaput N, André F, Caby MP, Novault S, Flament C, Leboulaire C, Borg C, Amigorena S, Boccaccio C, Bonnerot C, Dhellin O, Movassagh M, Piperno S, Robert C, Serra V, Valente N, Le PJB, Spatz A, Lantz O, Tursz T, Angevin E, Zitvogel L (2005) Vaccination of metastatic melanoma patients with autologous dendritic cell (DC) derived-exosomes: results of thefirst phase I clinical trial. *J Transl Med* 3: 10.

[33] Lobb RJ, Becker M, Wen SW, Wong CSF, Wiegmans AP, Leimgruber A, Möller A (2015) Optimized exosome isolation protocol for cell culture supernatant and human plasma. *J Extracell Vesicles* 4: 27031. https://doi.org/10.3402/jev.v4.27031.

[34] Greening DW, Xu R, Ji H, Tauro BJ, Simpson RJ (2015) A protocol for exosome isolation and characterization: evaluation of ultracentrifugation, density-gradient separation, and immunoaffinity capture methods. *Methods Mol Biol* 1295: 179–209.

[35] Kalra H, Adda CG, Liem M, Ang CS, Mechler A, Simpson RJ, Hulett MD, Mathivanan S (2013) Comparative proteomics evaluation of plasma exosome isolation techniques and assessment of the stability of exosomes in normal human blood plasma. *Proteomics* 13: 3354–3364.

[36] Witwer KW, Buzás EI, Bemis LT, Bora A, Lässer C, Lötvall J, Noltet HEN, Piper MG, Sivaraman S, Skog J, Théry C, Wauben MH, Hochberg F (2013) Standardization of sample collection, isolation and analysis methods in extracellular vesicle research. *J Extracell Vesicles* 2. https://doi.org/10.3402/jev.v2i0.20360.

[37] Tominaga N, Yoshioka Y, Ochiya T (2015) A novel platform for cancer therapy using extracellular vesicles. *Adv Drug Deliv Rev* 95: 50–55.

[38] Nordin JZ, Lee Y, Vader P, Mäger I, Johansson HJ, Heusermann W, Wiklander OPB, Hällbrink M, Seow Y, Bultema JJ, Gilthorpe J, Davies T, Fairchild PJ, Gabrielsson S, Meisner-kober NC, Lehtiö J, Smith CIE, Wood MJA, Andaloussi SEL (2015) Ultrafiltration with size-exclusion liquid chromatography for high yield isolation of extracellular vesicles preserving intact biophysical and functional properties. Nanomedicine Nanotechnology, *Biol Med* 11: 879–883.

[39] Kosaka N, Yoshioka Y, Hagiwara K, Tominaga N, Ochiya T (2013) Functional analysis of exosomal microRNA in cell-cell communication research. *Methods Mol Biol* 1024: 1–10.

[40] Christ L, Raiborg C, Wenzel EM, Campsteijn C, Stenmark H (2017) Cellular functions and molecular mechanisms of the ESCRT membrane-scission machinery. *Trends Biochem Sci* 42: 42–56.

[41] Klumperman J, Raposo G (2014) The complex ultrastructure of the endolysosomal system. *Cold Spring Harb Perspect Biol* 6: a016857.

[42] Hurley JH (2008) ESCRT complexes and the biogenesis of multivesicular bodies. *Curr Opin Cell Biol* 20: 4–11.

[43] Baietti MF, Zhang Z, Mortier E, Melchior A, Degeest G, Geeraerts A, Ivarsson Y, Depoortere F, Coomans C, Vermeiren E, Zimmermann P, David G (2012) Syndecan-syntenin-ALIX regulates the biogenesis of exosomes. *Nat Cell Biol* 14: 677–685.

[44] Morita E, Sundquist WI (2004) Retrovirus budding. *Annu Rev Cell Dev Biol* 20: 395–425.

[45] Webster BM, Colombi P, Jäger J, Lusk CP (2014) Surveillance of nuclear pore complex assembly by ESCRT-III/Vps4. *Cell* 159: 388–401.

[46] Olmos Y, Hodgson L, Mantell J, Verkade P, Carlton JG (2015) ESCRT-III controls nuclear envelope reformation. *Nature* 522: 236–239.

[47] Vietri M, Schink KO, Campsteijn C, Wegner CS, Schultz SW, Christ L, Thoresen SB, Brech A, Raiborg C, Stenmark H (2015) Spastin and

ESCRT-III coordinate mitotic spindle disassembly and nuclear envelope sealing. *Nature* 522: 231–235.

[48] Buschow SI, Hoen ENM, Nolte T, Niel G van, Pols MS, Broeke TT, Lauwen M, Ossendorp F, Melief CJM, Raposo G, Wubbolts R, Wauben MHM, Stoorvogel W (2009) MHC II in dendritic cells is targeted to lysosomes or T cell-induced exosomes via distinct multivesicular body pathways. *Traffic* 10: 1528–1542.

[49] Jimenez AJ, Maiuri P, Lafaurie-Janvore J, Divoux S, Piel M, Perez F (2014) ESCRT machinery is required for plasma membrane repair. *Science* 343: 1247136.

[50] Theos AC, Truschel ST, Tenza D, Hurbain I, Harper DC, Berson JF, Thomas PC, Raposo G, Marks MS (2006) A lumenal domain-dependent pathway for sorting to intralumenal vesicles of multivesicular endosomes involved in organelle morphogenesis. *Dev Cell* 10: 343–354.

[51] Niel G van, Charrin S, Simoes S, Romao M, Rochin L, Saftig P, Marks MS, Rubinstein E, Raposo G (2011) The tetraspanin CD63 regulates ESCRT-independent and -dependent endosomal sorting during melanogenesis. *Dev Cell* 21: 708–721.

[52] Trajkovic K, Hsu C, Chiantia S, Rajendran L, Wenzel D, Wieland F, Schwille P, Brügger B, Simons M (2008) Ceramide triggers budding of exosome vesicles into multivesicular endosomes. *Science* 319 1244–1247.

[53] Kosaka N, Iguchi H, Hagiwara K, Yoshioka Y, Takeshita F, Ochiya T (2013) Neutral sphingomyelinase 2 (nSMase2)-dependent exosomal transfer of angiogenic micrornas regulate cancer cell metastasis. *J Biol Chem* 288: 10849–10859.

[54] Chairoungdua A, Smith DL, Pochard P, Hull M, Caplan MJ (2010) Exosome release of β-catenin: a novel mechanism that antagonizes Wnt signaling. *J Cell Biol* 190: 1079–1091.

[55] Rao J, Fitzpatrick RE (2004) Use of the Q-switched 755-nm alexandrite laser to treat recalcitrant pigment after depigmentation therapy for vitiligo. *Dermatol Surg* 30: 1043–1045.

[56] Liégeois S, Benedetto A, Garnier J-M, Schwab Y, Labouesse M (2006) The V0-ATPase mediates apical secretion of exosomes containing Hedgehog-related proteins in Caenorhabditis elegans. *J Cell Biol* 173: 949–961.

[57] Tricarico C, Clancy J, D'Souza-Schorey C (2017) Biology and biogenesis of shed microvesicles. *Small GTPases* 8: 220–232.

[58] Villarroya-Beltri C, Baixauli F, Mittelbrunn M, Fernández-Delgado I, Torralba D, Moreno-Gonzalo O, Baldanta S, Enrich C, Guerra S, Sánchez-Madrid F (2016) ISGylation controls exosome secretion by promoting lysosomal degradation of MVB proteins. *Nat Commun* 7: 13588. https://doi.org/10.1038/ncomms13588.

[59] Edgar JR, Manna PT, Nishimura S, Banting G, Robinson MS (2016) Tetherin is an exosomal tether. *Elife* 5. https://doi.org/10.7554/eLife.17180.

[60] Dias MVS, Teixeira BL, Rodrigues BR, Sinigaglia-Coimbra R, Porto-Carreiro I, Roffé M, Hajj GNM, Martins VR (2016) PRNP/prion protein regulates the secretion of exosomes modulating CAV1/caveolin-1-suppressed autophagy. *Autophagy* 12: 2113–2128.

[61] Hessvik NP, Øverbye A, Brech A, Torgersen ML, Jakobsen IS, Sandvig K, Llorente A (2016) PIKfyve inhibition increases exosome release and induces secretory autophagy. *Cell Mol Life Sci* 73: 4717–4737.

[62] Bonifacino JS, Glick BS (2004) The mechanisms of vesicle budding and fusion. *Cell* 116: 153–166.

[63] Cai H, Reinisch K, Ferro-Novick S (2007) Coats, tethers, Rabs, and SNAREs work together to mediate the intracellular destination of a transport vesicle. *Dev Cell* 12: 671–682.

[64] Hsu C, Morohashi Y, Yoshimura S-I, Manrique-Hoyos N, Jung S, Lauterbach MA, Bakhti M, Grønborg M, Möbius W, Rhee J, Barr FA, Simons M (2010) Regulation of exosome secretion by Rab35 and its GTPase-activating proteins TBC1D10A-C. *J Cell Biol* 189: 223–232.

[65] Savina A, Furlán M, Vidal M, Colombo MI (2003) Exosome release is regulated by a calcium-dependent mechanism in K562 cells. *J Biol Chem* 278: 20083–20090.

[66] Théry C, Zitvogel L, Amigorena S (2002) Exosomes: composition, biogenesis and function. *Nat Rev Immunol* 2: 569–579.

[67] Abels ER, Breakefield XO (2016) Introduction to extracellular vesicles: Biogenesis RNA cargo selection, content, release, and uptake. *Cell Mol Neurobiol* 36: 301–312.

[68] Katsuda T, Kosaka N, Takeshita F, Ochiya T (2013) The therapeutic potential of mesenchymal stem cell-derived extracellular vesicles. *Proteomics.* 13: 1637–1653.

[69] Cocucci E, Meldolesi J (2015) Ectosomes and exosomes: shedding the confusion between extracellular vesicles. *Trends Cell Biol* 25: 364–372.

[70] Frühbeis C, Fröhlich D, Kuo WP, Amphornrat J, Thilemann S, Saab AS, Kirchhoff F, Möbius W, Goebbels S, Nave K-A, Schneider A, Simons M, Klugmann M, Trotter J, Krämer-Albers E-M (2013) Neurotransmitter-triggered transfer of exosomes mediates oligodendrocyte-neuron communication., *PLoS Biol* 11: e1001604. https://doi.org/10.1371/journal.pbio.1001604.

[71] Lopez-Verrilli MA, Picou F, Court FA (2013) Schwann cell-derived exosomes enhance axonal regeneration in the peripheral nervous system. *Glia* 61: 1795–1806.

[72] Barraud-Lange V, Naud-Barriant N, Bomsel M, Wolf J-P, Ziyyat A (2007). Transfer of oocyte membrane fragments to fertilizing spermatozoa. *FASEB J* 21: 3446–3449.

[73] Tannetta D, Dragovic R, Alyahyaei Z, Southcombe J (2014) Extracellular vesicles and reproduction-promotion of successful pregnancy. *Cell Mol Immunol* 11: 548–563.

[74] Peinado H, Alečković M, Lavotshkin S, Matei I, Costa-Silva B, Moreno-Bueno G, Hergueta-Redondo M, Williams C, García-Santos G, Ghajar CM, Nitadori-Hoshino A, Hoffman C, Badal K, Garcia BA, Callahan MK, Yuan J, Martins VR, Skog J, Kaplan RN, Brady MS, Wolchok JD, Chapman PB, Kang Y, Bromberg J, Lyden D

(2012) Melanoma exosomes educate bone marrow progenitor cells toward a pro-metastatic phenotype through MET, *Nat Med* 18: 883–891.

[75] Tominaga N, Kosaka N, Ono M., Katsuda T., Yoshioka Y., Tamura K., Lötvall J., Nakagama H., Ochiya T. (2015) Brain metastatic cancer cells release microRNA-181c-containing extracellular vesicles capable of destructing blood-brain barrier., *Nat. Commun.* 6: 6716. https://doi.org/10.1038/ncomms7716.

[76] Tominaga N, Katsuda T, Ochiya T (2015) Micromanaging of tumor metastasis by extracellular vesicles. *Semin Cell Dev Biol* 40: 52–59.

[77] Fujita Y, Yoshioka Y, Ito S, Araya J, Kuwano K, Ochiya T (2014) Intercellular communication by extracellular vesicles and their microRNAs in asthma. *Clin Ther* 36: 873–881.

[78] Gupta S, Knowlton AA (2007) HSP60 trafficking in adult cardiac myocytes: role of the exosomal pathway. *Am J Physiol Heart Circ Physiol* 292: H3052-H3056.

[79] Schorey JS, Harding CV (2016) Extracellular vesicles and infectious diseases: new complexity to an old story. *J Clin Invest* 126: 1181–1189.

[80] Fevrier B, Vilette D, Archer F, Loew D, Faigle W, Vidal M, Laude H, Raposo G (2004) Cells release prions in association with exosomes. *Proc Natl Acad Sci USA* 101: 9683–9688. https://doi.org/10.1073/pnas.0308413101.

[81] Guay C, Menoud V, Rome S, Regazzi R (2015) Horizontal transfer of exosomal microRNAs transduce apoptotic signals between pancreatic beta-cells. *Cell Commun Signal* 13: 17.

[82] Pisitkun T, Shen R-F, Knepper MA (2004) Identification and proteomic profiling of exosomes in human urine. *Proc Natl Acad Sci USA* 101: 13368–13373. https://doi.org/10.1073/pnas.0403453101.

[83] Cai Z, Yang F, Yu L, Yu Z, Jiang L, Wang Q, Yang Y, Wang L, Cao X, Wang J (2012) Activated T cell exosomes promote tumor invasion via Fas signaling pathway. *J Immunol* 188: 5954–5961.

[84] Saunderson SC, Dunn AC, Crocker PR, McLellan AD (2014) CD169 mediates the capture of exosomes in spleen and lymph node. *Blood* 123: 208–216.

[85] Zhang B, Wang M, Gong A, Zhang X, Wu X, Zhu Y, Shi H, Wu L, Zhu W, Qian H, Xu W (2015) HucMSC-exosome mediated-Wnt4 signaling is required for cutaneous wound healing. *Stem Cells* 33: 2158–2168.

[86] Zhu Q, Kang J, Miao H, Feng Y, Xiao L, Hu Z, Liao D-F, Huang Y, Jin J, He S (2014) Low-dose cytokine-induced neutral ceramidase secretion from INS-1 cells via exosomes and its anti-apoptotic effect. *FEBS J* 281: 2861–2870.

[87] Santiago-Dieppa DR, Steinberg J, Gonda D, Cheung VJ, Carter BS, Chen CC (2014) Extracellular vesicles as a platform for "liquid biopsy" in glioblastoma patients. *Expert Rev Mol Diagn* 14: 819–825.

[88] Yoshioka Y, Kosaka N, Konishi Y, Ohta H, Okamoto H, Sonoda H, Nonaka R, Yamamoto H, Ishii H, Mori M, Furuta K, Nakajima T, Hayashi H, Sugisaki H, Higashimoto H, Kato T, Takeshita F, Ochiya T (2014) Ultra-sensitive liquid biopsy of circulating extracellular vesicles using ExoScreen. *Nat. Commun* 5: 3591. https://doi.org/10.1038/ncomms4591.

[89] Lee H, Han KH, Lee SE, Kim SH, Kang HG, Cheong H Il (2012) Urinary exosomal WT1 in childhood nephrotic syndrome. *Pediatr Nephrol* 27: 317–320.

[90] Ichii O, Otsuka-Kanazawa S, Horino T, Kimura J, Nakamura T, Matsumoto M, Toi M, Kon Y (2014) Decreased miR-26a expression correlates with the progression of podocyte injury in autoimmune glomerulonephritis. *PLoS One* 9: e110383. https://doi.org/10.1371/journal.pone.0110383.

[91] Solé C, Cortés-Hernández J, Felip ML, Vidal M, Ordi-Ros J (2015) miR-29c in urinary exosomes as predictor of early renal fibrosis in lupus nephritis. *Nephrol Dial Transplant* 30: 1488–1496.

[92] Zubiri I, Posada-Ayala M, Sanz-Maroto A, Calvo E, Martin-Lorenzo M, Gonzalez-Calero L, Cuesta F de la, Lopez JA, Fernandez-

Fernandez B, Ortiz A, Vivanco F, Alvarez-Llamas G (2014) Diabetic nephropathy induces changes in the proteome of human urinary exosomes as revealed by label-free comparative analysis. *J Proteomics* 96: 92–102.

[93] Wang D, Sun W (2014) Urinary extracellular microvesicles: isolation methods and prospects for urinary proteome. *Proteomics* 14: 1922–1932.

[94] Barutta F, Tricarico M, Corbelli A, Annaratone L, Pinach S, Grimaldi S, Bruno G, Cimino D, Taverna D, Deregibus MC, Rastaldi MP, Perin PC, Gruden G (2013) Urinary exosomal microRNAs in incipient diabetic nephropathy., *PLoS One.* 8: e73798. https://doi.org/10.1371/journal.pone.0073798.

[95] Eissa S, Matboli M, Aboushahba R, Bekhet MM, Soliman Y (2016) Urinary exosomal microRNA panel unravels novel biomarkers for diagnosis of type 2 diabetic kidney disease. *J Diabetes Complications* 30: 1585–1592.

[96] Eissa S, Matboli M, Bekhet MM (2016) Clinical verification of a novel urinary microRNA panal: 133b, -342 and -30 as biomarkers for diabetic nephropathy identified by bioinformatics analysis. *Biomed Pharmacother* 83: 92–99.

[97] Zubiri I, Posada-Ayala M, Benito-Martin A, Maroto AS, Martin-Lorenzo M, Cannata-Ortiz P, Cuesta F de la, Gonzalez-Calero L, Barderas MG, Fernandez-Fernandez B, Ortiz A, Vivanco F, Alvarez-Llamas G (2015) Kidney tissue proteomics reveals regucalcin downregulation in response to diabetic nephropathy with reflection in urinary exosomes. *Transl Res* 166: 474-484.e4.

ABOUT THE EDITOR

Masayoshi Yamaguchi, PhD
Visiting Professor, Laboratory of Pathology and Laboratory Medicine,
David Geffen School of Medicine, University of California,
Los Angeles (UCLA), Los Angeles, CA, US

Masayoshi Yamaguchi, PhD, IOM, FAOE, DDG, DG, is Visiting Professor, Department of Pathology and Laboratory Medicine, David Geffen School of Medicine, University of California, Los Angeles (UCLA) (2017-2019); Adjunct Professor, Department of Hematology and Medical Oncology, Emory University School of Medicine, Atlanta, GA, USA (2013-2016); Visiting Professor, Department of Medicine, Baylor College of Medicine, Houston, USA (2012-2013); Visiting Professor, Department of Medicine, Emory University School of Medicine (2007-2011); Full Professor, Graduate School of Nutritional Sciences, University of Shizuoka, Shizuoka, Japan (1992-2007). Dr. Yamaguchi is engaged in the fields of endocrinology and cell signaling since 1971, and these areas of study are developed in the aspects of biochemistry, molecular biology, endocrinology, metabolism, nutrition, pharmacology and toxicology. Dr. Yamaguchi discovered two novel proteins (genes), which were named a regucalcin, a cell signaling suppressor (1978), and RGPR-p117, a regucalcin gene promoter-related transcription factor (2001). Dr. Yamaguchi generated regucalcin transgenic rats, which were registered as international patents including USA, EU and Japan, and this animal model was found to induce osteoporosis and hyperlipidemia. Moreover, Dr. Yamaguchi proposed the potential role of regucalcin as a novel suppressor in human carcinogenesis. Since 1974, Dr. Yamaguchi published over 550 English articles in professional journals with peer-review, and registered 23 national and international patents. Dr. Yamaguchi serviced as Editorial Board Members in over 80 Journals. Dr. Yamaguchi was listed in many biographies books. Dr. Yamaguchi received "The 2017 Albert Nelson Marquis Lifetime Achievement Award"（Marquis Who's Who, USA）.

INDEX

#

17β-estradiol, 36, 64, 65, 83, 87
25(OH)$_2$D$_3$, 82, 83, 84, 85, 87
25-vitamin D$_3$, 82
3T3-L1 adipocytes, 92, 93, 101
5α-dihydrotestosterone, 21
6-dichloro-1-β-D-ribofuranosylbenzimidazole (DRB), 91
6-diphosphatase, 111, 115

β

β-catenin, 65, 69, 70, 73, 139

λ

λZAP II, 2

A

AAbs, 52, 54, 61
accessory glands, 16
acetyl-CoA carboxylase, 107
acrosomal reaction, 24, 30, 49
acrosomal region, 24, 29, 33
acrosome, 23, 24, 30, 32, 34, 47, 49
acrosome reaction, 30, 33, 34, 47, 49
actinomycin D, 28, 45, 71, 72
adipocytes, 92, 93, 96, 97, 105
adipocytokines, 111
adipogenesis, x, 80, 92, 93, 94, 96, 97, 101, 105, 109, 116
adiponectin, 111, 112
AFP, 51, 52, 59, 60, 73
AFP-negative liver cancer serum, 51, 73
aging, viii, 16, 29, 36, 38, 46, 47, 59, 78, 85, 86, 95, 96, 97, 117, 119, 134
AKT, 71, 72
alanine transaminase, 28
albumin, 96, 110
ALG-2 interacting protein X, 128
ALIX, 128, 130, 138
alkaline phosphatase, 88, 89, 91, 92, 95, 96, 100
alpha-1-microglobulin/bikunin precursor, 133
alpha-fetoprotein, 52, 59
alternative splicing, viii, 1, 4
androgen receptor, 19, 22, 41, 42
androgens, 20, 21, 22, 42, 43, 44
antioxidant, 17, 27, 29, 119

anti-regucalcin antibody, ix, 51, 52, 57, 58, 61, 73
anti-RGN antiserum, 3
AP-1, 65
apoE, 112
apoE-deficient mice, 112
apolipoprotein C1 TG, 111
apolipoprotein C1 TG mice, 111
apolipoprotein C3, 111
apolipoprotein C3-KO mice, 111
apoptosis, viii, ix, x, 16, 17, 20, 21, 22, 25, 27, 28, 31, 33, 38, 42, 43, 44, 45, 52, 71, 72, 73, 77, 78, 79, 81, 88, 104, 131, 132
artificial insemination, 31
ascorbic acid, 13
ATG codon, 3
autoimmune response, 53
autophagosome, 130
autophagy, 130, 140

B

B cells, 131
bacteria, 17
Bax, 28, 71, 72
Bcl-2, 28, 71, 72, 101
B-lymphocytes, 61, 136
bone, v, x, 70, 76, 79, 80, 81, 82, 83, 84, 85, 86, 87, 88, 89, 92, 93, 94, 95, 96, 97, 98, 99, 100, 101, 102, 105, 109, 110, 116, 136, 142, 145
bone formation, 80, 81, 86, 87, 88, 89, 95, 100
bone homeostasis, x, 79, 80, 93, 94, 96, 97, 98
bone loss, x, 79, 80, 83, 85, 86, 92, 95, 96, 97, 98, 99, 101, 102, 110, 116
bone marrow, x, 79, 80, 82, 83, 84, 85, 86, 92, 93, 94, 96, 97, 98, 99, 101, 105, 109, 116, 136, 142

bone marrow cells, x, 79, 82, 84, 85, 86, 92, 94, 97, 98, 105, 109
bone marrow mesenchymal stem cell (MSC), 86, 92, 101
bone mass, 80, 82, 86, 95
bone metabolism, 80, 98, 116
bone remodeling, v, x, 79, 80, 81, 82, 94, 98
bone resorption, 80, 82, 83, 85, 86, 87, 97, 98
bovine, 4, 21, 22, 23, 31, 43, 44, 47, 71, 87, 126
breast, xi, 19, 40, 54, 58, 62, 63, 64, 68, 70, 76, 78
breast cancer cells, 70
buffalo, 23, 24, 30, 31, 32, 37, 39, 48
buffalo sperm, 24, 32
bulbourethral glands, 21

C

Ca^{2+}, vii, viii, ix, 1, 2, 6, 9, 10, 11, 12, 13, 15, 16, 17, 18, 22, 24, 25, 27, 31, 32, 33, 34, 35, 37, 39, 40, 44, 45, 48, 49, 50, 65, 70, 71, 72, 88, 90, 100, 105, 106, 111, 115
Ca^{2+} absorption, 34
Ca^{2+} binding constant, 2
Ca^{2+} binding protein, 9
Ca^{2+}- efflux, 32
Ca^{2+} handling, 22, 34
Ca^{2+} homeostasis, viii, ix, 15, 16, 17, 32
Ca^{2+} influx, 27, 33, 44
Ca^{2+}/calmodulin, 2, 12, 70, 88, 90
Ca^{2+}/calmodulin-dependent protein tyrosine phosphatase, 70
Ca^{2+}/Mg^{2+}-ATPase, 32
Ca^{2+}-ATPases, 32
Ca^{2+}-binding ability, 34
Ca^{2+}-binding protein, vii, 2, 9, 11, 13, 25, 35, 37
Ca^{2+}-binding site, 9

Ca^{2+}-permeable channels, 34
Ca^{2+}-pumps, 16
Ca^{2+}stress, 32
cadmium chloride, 27
calbindin-D28k, 9
calcitonin, 64, 65, 82, 83
calcitonin and 1, 82
calcium, vii, ix, 2, 9, 10, 11, 12, 13, 15, 16, 35, 36, 37, 39, 40, 43, 44, 47, 48, 49, 50, 64, 79, 81, 83, 85, 95, 96, 99, 100, 104, 105, 110, 114, 115, 116, 117, 118, 119, 141
calcium-binding domain, 104
calcium-binding protein, 2, 11, 12, 13, 35, 36, 39, 43, 48, 79, 104, 114, 115, 116, 117, 118, 119
calmodulin, vii, viii, 9, 13, 65, 88, 111, 115
CaM kinase, 65
CaMKP, 70
cancellous bone, 81
cancer metastasis, 132
capacitation, 24, 30, 31, 33, 34
carbon tetrachloride, 64, 65, 112, 118
carbonic anhydrase-7, 106
carcinogenesis, 12, 38, 39, 43, 62, 63, 66, 77, 145
cardiolipin, 110
caspase-3, 28, 71, 72
caspase-8, 28, 71, 72
caspases, 28
catalase, 29, 47
catalase and glutathione peroxidase, 29
cBioPortal, 61, 66
CD63, 130, 139
CD81, 130
CD82, 130
CD9, 130
cdc2, 70
cdk 5, 70
cDNA, viii, 1, 2, 3, 4, 6, 12, 35, 36, 41, 43, 52, 55, 57, 114
cDNA and genomic cloning, 2

cDNA probe, 4
cell death, x, 17, 25, 27, 52, 70, 71, 72, 77, 101
cell-division protein kinase-4, 106
cells, vii, viii, ix, x, xi, 10, 20, 21, 23, 24, 27, 28, 32, 33, 40, 42, 44, 45, 52, 54, 62, 65, 68, 70, 71, 72, 73, 76, 77, 78, 79, 80, 81, 82, 83, 84, 85, 86, 87, 88, 89, 90, 91, 92, 93, 94, 95, 96, 97, 98, 100, 101, 103, 104, 105, 106, 107, 108, 109, 114, 116, 121, 122, 123, 124, 126, 131, 132, 133, 134, 136, 141, 142, 143
cellular immunity, 54
c-fos, 70
ChIP assays, 67
chk2m, 70
cholesterol, 30, 47, 96, 110, 111, 118
cholesterol 7 alpha-hydroxylase, 112
cholesterol 7 alpha-hydroxylase-deficient mice, 112
chromosomal localization, 3
chronic glomerulonephritis, 132
circular dichroism, 18, 39
c-Jun, 70
c-kit, 42, 70
Clathrin, 128
c-myc, 70
cold tolerance, 32, 48, 104, 115
collagen, 88, 96
complementary DNA, 2
CpG methylation, 66
CpG sequence, 66, 67, 68, 73
cryopreservation, 16, 29, 31, 32, 33, 34, 35, 47, 48, 50
crystal, 9, 10, 13, 18, 36
crystal structure, 9, 18
c-src, 70
CT, 53, 65
cytochrome C, 71, 72
cytokine, x, 80, 83, 86, 97, 98, 132, 143
cytoplasm, 18, 19, 23, 32, 65, 70, 71, 73, 111, 113, 129

D

DAC, 67, 68
decapacitation factors, 30
delta/jagged proteins, 92
dendritic cell, 124, 131, 136, 137, 139
DHT, 21, 65
diabetes, v, x, 103, 104, 105, 106, 108, 109, 114, 115, 117, 118, 131, 132, 133, 144
diabetic nephropathy, xi, 121, 132, 133, 144
diagnostic biomarker, 54
dissociation constant, 9
disulfide isomerase A6, 106
divalent metal, 9
DNA, 6, 13, 22, 37, 41, 45, 57, 64, 66, 67, 68, 71, 87, 90, 91, 95, 104
DNA fragmentation, 71
DNA methylation, 64, 66
DNA methyltransferase inhibitor, 67, 68
DRB, 91
Drosophila montana, 32, 48, 104
ducts fluids, 16
dynein, 130

E

E 3 ubiquitin protein ligase, 106
EBV, 54
E-cadherin, 69
ectosome, 122
EF-hand, vii, 9, 18, 104
EF-hand motif, vii, 9, 104
elevate glucose transporter 2 (GLUT 2) mRNA, 107
ELISA, 57, 58
EMBL, 3
epididymal lumen, 26, 33
epididymis, 23, 24, 26, 29, 33, 37, 45
erythrocyte, 123, 134
ESCRT, 128, 129, 130, 138, 139
ESCRT-0, 128, 129
ESCRT-I, 128, 130, 138, 139
ESCRT-III, 128, 130, 138, 139
estrogen, 20, 21, 22, 25, 42, 43, 64, 95, 97, 99, 101
estrogen receptors, 20, 42, 43
ethanol, 112, 116, 119
eukaryotes, ix, 15, 16, 18
eukaryotic translation initiation factor-3, 106
evolution, ix, 9, 15, 16, 18
EVs, 122, 124, 125, 126, 127, 130, 131, 132, 133
excurrent, 16, 30
exogenous regucalcin, 77, 88, 89, 90, 100, 109
exon, 3, 4, 5, 13, 19
exon 2, 5
exosome, xi, 122, 123, 124, 134, 135, 137, 139, 140, 141, 143
extracellular regucalcin, x, 79, 83, 84, 85, 88, 89, 90, 91, 92, 93, 94, 97
extracellular signal-regulated kinase (ERK), 37, 38, 70, 92, 93, 101
extracellular vesicles, vi, xi, 121, 122, 123, 125, 127, 128, 130, 131, 132, 134, 138, 141, 142, 143

F

FAS, 131
FASL, 131
fasting, 105
fatty acid synthase, 131
fatty acid synthase ligand, 131
fertility, 19, 31, 32, 33, 34
fibroblastic growth factors, 92
fibrosis, xi, 104, 113, 143
FISH, 3
fluids, 16, 24, 30, 33, 89
fructose 1, 111, 115
fungi, 17

G

galactosamine, 112, 119
GenBank, 3
genomic conservation, 6
germ cell maturation, 20
germ cells, 20, 22, 23, 25, 27, 28, 33
glioblastoma, 56, 143
glomerulosclerosis, 132
glucokinase, 107
gluconolactonase, viii, 2, 5, 6, 7, 8, 10, 13, 18, 29, 39, 102, 116, 117
glucose, x, 16, 28, 46, 83, 103, 104, 105, 106, 107, 108, 109, 110, 111, 114, 115, 116, 117, 118
glucose and lipid metabolism, 105, 106, 107, 114
glucose levels, 16, 108
glucose metabolism, 104, 105, 106, 118
glucose or urea nitrogen, 110
glucose supplementation, x, 103, 106, 107
glucose tolerance, xi, 103, 108, 109, 116, 117
glucose transporter 2, xi, 104
GLUT 2 mRNA, 107
glutamine, 28, 46
glutathione peroxidase, 29, 30
glutathione S-transferase, 27
glyceroaldehyde-3-phosphate dehydrogenase, 88, 107
glyceroaldehyde-3-phosphate dehydrogenase (G3PDH), 88, 107
glyceroaldehyde-3-phosphate dehydrogenase (G3PDH) mRNAs, 107
glycogen, xi, 104, 111, 115
glycogen content, xi, 104, 111
glycogenesis, 105
glycogenolysis, 111
glycolysis, 105
glycolytic enzyme, 111
glycolytic metabolism, 17, 39, 111, 117

gonadal hormones, 20
gonads, 16, 20
GPI, 124
G-proteins, 106
growth factors, 28, 82
Guangxi HCC cDNA library screening, 56

H

H_2O_2, 71
H4-II-E cells, 106, 107
HCC-22-5, 56, 74
HCC-associated antigen, 56, 57
HDL-cholesterol, 96, 110
heart and brain, 27, 83
hedgehogs, 92
helix, 9, 10
helix-loop-helix domain, 9
hematopoietic progenitors, 81, 82
hepatic stellate cells (HSCs), 113, 119
hepatitis B&C, 56
hepatocellular carcinoma, 52, 55, 59, 61, 63, 67, 68, 74, 75, 76
hepatoma H4-II-E cells, xi, 70, 77, 103, 106, 109, 116
Her2-Neu, 54
high-density lipoprotein (HDL), 96, 110
histone acetylation, 66
histone deacetylase inhibitor, 66
histone lysine N-methyltransferase, 133
HMG-CoA reductase, 107
homology, viii, 1, 3, 6, 7, 8, 9
HP, 24, 25
HPV, 54
H-ras, 70, 73
Huh7, 66
human breast cancer cells, 72, 76
humoral immunity, 54
hydrophility, 9
hyperglycolytic profile, 111

hyperlipidemia, x, 39, 92, 96, 104, 105, 109, 110, 111, 112, 114, 115, 117, 118, 145
hypospermatognesis, 24
hypothalamic, 20

I

IGF- I, 82
immune cell, 61, 131
immune response, 52, 54, 56, 57, 61, 74, 124, 125, 132, 136
immunogenicity, 61
in situ hybridization, 3, 62
infertile men, 25
inflammation, 53, 54, 61, 132
inorganic phosphorus, 96
insect, 35, 36, 48, 104, 115
insulin, x, 16, 64, 71, 82, 92, 93, 101, 103, 105, 106, 107, 108, 109, 113, 114, 116, 117, 118
insulin receptor, xi, 93, 104, 107
insulin resistance, x, 103, 105, 106, 107, 113, 114, 116
insulin signaling, xi, 104, 107, 116
insulin-like growth factor, 71, 82, 101
intracellular signaling, vii, viii, ix, 15, 17, 69, 71, 77, 83, 87, 89, 90, 91, 93, 100
intracellular signaling factors, 77, 87, 89, 90, 91
intron, 3, 13
invasion, x, 52, 53, 59, 63, 69, 70, 73, 142
invertebrate species, 16
Islet, 109
isoform, 4, 6
isoforms, viii, 1, 4, 5, 6, 24, 42

K

Kaplan-Meier analyses, 64
Kd, 9

kidney, ix, 12, 15, 16, 21, 40, 41, 43, 45, 51, 62, 63, 70, 73, 77, 83, 131, 132, 144
kidney disease, 131, 132, 144
kinases, 17, 71, 87, 90
kinesins, 130
kinogen heavy chain, 106
knockdown, 67, 68

L

lactate, 28, 33, 46, 111
lactate dehydrogenase, 111
lactic acid, 83
L-ascorbic acid, 18, 29, 39, 102, 108
LDL receptor, 118
leptin, xi, 104, 111, 112, 117
leptin and adiponectin, xi, 104, 111, 117
Leydig cells, 23, 24
lipid, x, 17, 27, 29, 93, 96, 103, 104, 105, 106, 107, 109, 110, 111, 112, 113, 114, 115, 116, 117, 123, 124, 131
lipid droplets, 109, 113
lipid metabolic disorder, xi, 104, 112, 114
lipid metabolism, 96, 104, 105, 106, 107, 109, 111, 112, 113, 115, 117
lipid peroxidation, 27, 29
lipid synthesis, 111
lipoprotein lipase, 111, 117, 118
lipoprotein lipase-deficient mice, 111
liquid biopsy, 132, 143
liver, ix, x, 2, 4, 5, 11, 12, 15, 16, 21, 27, 29, 36, 37, 38, 39, 40, 48, 51, 52, 54, 56, 57, 58, 60, 61, 62, 63, 65, 66, 70, 73, 75, 77, 78, 83, 90, 100, 103, 105, 106, 107, 108, 110, 111, 112, 113, 114, 115, 116, 117, 119
liver cancer, ix, 51, 52, 57, 58, 62, 65, 66, 70, 73, 75, 77, 105
liver cancer HepG2 cells, 70, 105
liver cirrhosis, 56, 63
liver disease, xi, 104, 112, 119

liver myr-Akt overexpressing mice, 112
LNCaP prostate cancer cells, 22, 44
low-density lipoprotein (LDL), 111, 113, 118
LPS, 65, 71
lysosome, 130

M

macrophage, 83, 98, 131
macrophage colony-stimulating factor (M-CSF), 83, 84
MAGEA3, 53
magnesium, 18
male, v, ix, 2, 15, 16, 17, 18, 19, 20, 21, 22, 23, 24, 25, 27, 28, 29, 30, 33, 34, 36, 37, 39, 41, 42, 43, 46, 49, 85, 86, 95, 97, 110
male fertility, 16, 17, 19, 20, 21, 34, 41, 42, 49
mammalian, viii, ix, 1, 6, 8, 15, 17, 18, 19, 20, 24, 29, 31, 37, 42, 43, 45, 49, 50, 135
mammary gland, 21, 36, 38, 72
manganese, 18
MAPK kinase, 65
maturation, 20, 24, 26, 33, 36, 37, 44, 45, 124, 127, 134, 135, 136
MAWDBP, 106
MC3T3-E1 cell, 86, 88, 90
MCF-7, 21, 72, 78
MCF-7 breast cancer cells, 21
M-CSF, 83, 84
MDA-MB-231, 69, 70, 76
Melan-A/MART-1, 53
melanocytoma, 56
melanoma, 53, 137, 142
mesenchymal stem cell, 81, 92, 94, 96, 97, 101, 131, 141
mesoderm, 131
metabolic disorder, x, 39, 95, 103, 105, 112, 114, 115
metaphyseal tissues, 83, 85, 95

metastasis, 58, 59, 60, 76, 139, 142
methylation, ix, 52, 66, 67, 68, 73
methylation of regucalcin promoter, 66
MHCC97-H, 66
MIA PaCa-2, 69, 70, 77
mice, xi, 4, 6, 23, 24, 38, 39, 42, 46, 72, 95, 96, 99, 102, 103, 108, 109, 111, 113, 117, 118
microarray, 62
microRNA-26a, 132
microRNA-29c, 132
microvesicle, 122
mineral content, 95
mineral density, 95
mineral homeostasis, 80
mineralization, x, 80, 86, 88, 89, 90, 91, 92, 94, 95, 97, 99
miR-155, 133
miR-424, 133
missense mutation, 61
mitochondria, 16, 34, 109
mitogen-activated protein kinase (MAPK), 65, 70, 87, 90
modeling, 81, 96
monocarboxylate transporter 4, 111
MSCs, 92, 93
MUC-1, 53
multicellular unit, 81
multinucleated cell (MNC), 85
multivesicular bodies, xi, 121, 122, 124, 127, 130, 138
myosins, 130

N

nephritis, 132, 143
neutral lipids, 110, 117
neutral lipids and phospholipids including phosphatidylethanolamine, 110
NF1-A1, 65
NFI-A1, 65

NF-κB, x, 80, 84, 87, 91, 99
nitric oxide synthase, 40, 72
NLS, 18, 19
NO synthase, 71, 88
nonalcoholic fatty liver disease (NAFLD), 113
non-hodgkin's lymphoma, 56
non-translated region, 3
northern blotting, 4
NOS, 72
NRK52E cell, 40, 77, 78
NSCLC A549, 68, 70, 71
nuclear, viii, x, 5, 13, 18, 19, 20, 21, 34, 39, 40, 79, 82, 84, 91, 94, 104, 113, 129, 138, 139
nuclear factor kappa B ligand (RANKL), 82, 83, 84, 85, 97
nuclear localization signal, 18, 19
nucleus, 18, 19, 23, 32, 39, 40, 65, 69, 70, 91, 94
NY-ESO-1, 53

O

obesity, 104
obstructive azoospermia, 25
oligodendrocytes, 131
oncogene, x, 52, 62
oncosome, 122
oncoviral antigen, 54
oocyte, 26, 30, 131, 141
osteoblast, 92, 98, 100, 101
osteoblastic MC3T3-E1 cells, 86, 89, 92, 93, 99, 100
osteoblastogenesis, x, 80, 86, 88, 91, 92, 93, 94, 97, 99, 101, 115
osteoclast, x, 79, 81, 83, 84, 85, 86, 98, 99
osteoclastogenesis, x, 80, 82, 83, 84, 85, 86, 93, 94, 97
osteoporosis, v, x, 39, 79, 80, 86, 94, 95, 96, 97, 115, 145

osteoprotegerin (OPG), 84, 94
ovarian cancer, 56, 75
overall survival, 63, 64
overall survival rate, 63
overexpression, 28, 29, 38, 40, 45, 68, 70, 71, 72, 73, 76, 77, 94, 95, 96, 104, 106, 107, 111, 114, 116
oxidative damage, 27, 29, 31, 32, 33
oxidative stress, ix, 15, 16, 17, 27, 29, 33, 38, 43, 46, 47, 65, 71, 78

P

p21, 70
p53, 27, 70, 101
pancreatic β-cell, 109, 132
parathyroid hormone (PTH), 16, 64, 65, 82, 83, 84, 85, 87, 99
PBA, 66
PD98059, 87, 93
peripheral quantitative computed tomography (pQCT), 95
peroxisome proliferators-activated receptor-gamma (PPAR gamma), 92, 102, 113
p-GSK-3β, 70
phorbol 12-myristate 13-acetate, 67, 77
phosphatases, 17
phosphatidylcholine, 30, 110
phosphatidylinositide 3 (PI3), xi, 65, 90
phosphatidylinositol 3-kinase (PI3K), 70, 107
phosphatidylserine, 110
phosphatidylserine and sphingomyelin, 110
phosphodiesterase, 17
phosphofructokinase, 111
phospholipids, 117
phosphorylase a, 111, 115
phylogenetic tree analysis, 9
pituitary, 20
platelet-derived growth factor, 82
PMA, 67

polar strength strain index, 95
posttranslational modification, 55
post-translational modifications, 61
progenitor cell, 142
prognostic indicator, x, 52, 53, 76
prokaryotes, ix, 15, 16, 18
proliferation, viii, ix, x, 16, 17, 25, 31, 38, 39, 44, 52, 70, 71, 73, 75, 76, 77, 78, 79, 88, 91, 104, 117, 131
promoter region, 5, 66, 67, 68, 87
prostate, 19, 21, 23, 24, 29, 36, 38, 39, 40, 43, 44, 54, 62, 63, 72, 78, 111, 117
prostatic fluid, 124, 134
proteases, 17
proteasome subunit-alpha type 3, 106
protein disulfide isomerase A6, 106
protein kinase, vii, viii, x, 39, 40, 52, 65, 70, 71, 72, 73, 87, 90, 94
protein tyrosine kinase, 70
protein tyrosine phosphatase, 40, 70
proteins, vii, x, xi, 9, 18, 19, 30, 34, 35, 41, 47, 52, 53, 57, 61, 66, 71, 73, 92, 106, 107, 116, 121, 123, 124, 131, 135, 136, 140, 145
proteomics, 37, 62, 137, 141, 144
proton pump V-ATPase, 130
proto-oncogene, 62
PTMs, 55
PTP, 70
pyruvate kinase, 107, 111, 115

R

Rab GDP dissociation inhibitor-beta, 106
RAB11, 130
RAB27A, 130
RAB27B, 130
RAB35, 130
RANK, 83, 84, 99
Rat, 23, 25, 83
RAW267.4 cells, 84

Rb, 70
reactive oxygen species, 27, 32, 45, 46
redox equilibrium, 29
refeeding, 105, 116
regucalcin, v, vii, viii, ix, x, xi, xii, 1, 2, 11, 12, 13, 15, 16, 35, 36, 37, 38, 39, 40, 41, 43, 45, 46, 48, 51, 52, 53, 55, 56, 57, 58, 59, 60, 61, 62, 63, 64, 65, 66, 67, 68, 69, 70, 71, 72, 73, 75, 76, 77, 78, 79, 80, 82, 83, 84, 85, 86, 87, 88, 89, 90, 91, 92, 93, 94, 95, 96, 97, 98, 99, 100, 101, 103, 104, 105, 106, 107, 108, 109, 110, 111, 112, 113, 114, 115, 116, 117, 118, 119, 121, 122, 133, 144, 145
regucalcin autoantibody, 57, 59
regucalcin expression, 40, 62, 63, 64, 66, 67, 68, 72, 77, 112, 114
regucalcin gene, v, viii, 1, 12, 13, 36, 41, 48, 53, 57, 62, 64, 65, 66, 67, 72, 73, 76, 77, 82, 87, 95, 97, 99, 104, 105, 109, 114, 115, 145
regucalcin knockout mice, 96, 108, 109, 113
regucalcin mRNA, 12, 36, 43, 62, 64, 87, 94, 99, 105, 108, 109, 112, 116
Regucalcin mRNA, x, 73, 79, 86, 87, 105, 108
regucalcin protein, viii, 57, 61, 62, 63
regucalcin transgenic male and female rats, 85, 96, 110
regucalcin transgenic rat, x, xi, 39, 40, 46, 79, 80, 82, 83, 85, 86, 94, 95, 96, 97, 98, 99, 104, 105, 110, 111, 112, 115, 116, 117, 145
regucalcin transgenic rats, x, xi, 39, 40, 46, 79, 80, 82, 83, 85, 86, 94, 95, 96, 97, 98, 99, 104, 105, 110, 111, 112, 115, 116, 117, 145
regulator of G-protein signaling-5, 106
remodeling, 80, 81, 82
renal cell carcinoma, 56, 71, 76

reproduction, v, ix, 15, 16, 17, 19, 21, 22, 24, 25, 28, 31, 34, 37, 39, 42, 45, 46, 49, 141
reproductive tract, 16, 18, 22, 23, 24, 25, 29, 30, 31, 33, 37
RGN, 1, 2, 3, 4, 5, 6, 7, 8, 9, 10, 11, 15, 16, 17, 18, 19, 20, 21, 22, 23, 24, 25, 26, 27, 28, 29, 30, 31, 32, 33, 34, 35, 52, 53, 65, 67, 68, 69, 71, 72, 94
RGPRp117, 65
ribonuclease, 122
RNA, xi, 2, 39, 91, 104, 134, 141
runt-related transcription factor 2 (Runx2), 88, 89, 91, 92, 100
Runx2 mRNA, 88

S

S-100 protein, 9
S-100β, 9
SCF, 70
seminal fluid, 29
seminal vesicles, 24, 33
seminiferous tubules, 21, 22, 24, 27, 28, 33
senescence marker protein-30, 2, 12, 13, 16, 35, 36, 39, 41, 43, 45, 75, 117, 119
sequence alignment, 13, 17, 19, 42
SEREX, 52, 54, 55, 56, 57
SERPA, 54
Sertoli cell-only syndrome, 25
Sertoli cells, 23, 24, 28, 33, 42, 43, 46
sex-steroid target gene, 16
sheet, 9, 10
SK-HEP-1, 66, 67
Smad2/3, 113
SMP30, 2, 9, 10, 12, 16, 35, 36, 38, 39, 40, 41, 43, 46, 75, 76, 78, 102, 117
Southern hybridization, 6
SP1, 65, 66, 67, 77
sperm, ix, 16, 20, 22, 24, 26, 27, 28, 30, 31, 32, 33, 34, 35, 37, 43, 45, 47, 48, 49, 50

sperm capacitation, 16, 30, 37, 45, 47, 49
sperm counts, 26
sperm motility, 26, 27, 48
sperm parameters, 26, 27
sperm viability, 27, 28, 30
spermatogenesis, ix, 16, 17, 20, 21, 22, 24, 25, 28, 32, 33, 34, 42, 43, 44, 45, 46
spermatozoa, 16, 22, 23, 24, 29, 30, 31, 32, 33, 34, 37, 45, 47, 48, 49, 50, 141
sphingomyelin, 130
steroid hormone, 16, 20, 22, 40, 43
steroid receptors, 21, 87
streptozotocin (STZ), x, 103, 108, 116, 119
structural features of protein, 2
structure, v, viii, 1, 5, 7, 8, 9, 10, 11, 13, 17, 21, 35, 36, 42, 55, 75, 123
subunit 2, 106
superoxide dismutase, 29, 30, 38, 46, 106
survival, ix, 20, 22, 42, 52, 53, 63, 64, 71, 73, 76, 77, 78

T

T cell, 53, 131, 136, 139, 142
TAAs, 52, 53, 54, 61
tartrate-resistant acid phosphatase (TRACP), 85
t-BHP, 65, 71
tert-butyl hydroperoxide, 27
testicular, 20, 21, 24, 26, 27, 28, 33, 38, 42, 43, 45
testis, 19, 21, 22, 23, 24, 25, 28, 30, 37, 42, 43, 45, 53
testosterone, 21, 42, 43
TFAP2A, 66, 67, 68, 73
TFAP2A protein, 67, 68
TGF-β, 82, 89, 101
TGF-β1, 82, 89, 101
thapsigargin, 28, 45, 71
TNF receptor-associated factor (TRAF6), 87

total cholesterol, xi, 104, 110, 112
transcript, 4, 19
transcription factors, 21, 53, 65, 67, 73, 88
transforming growth factor, 27, 71, 82, 113
transforming growth factor-β, 27, 82, 113
transforming growth factor-β1 (TGF-β1), 82, 89, 101
transgenic, 25, 26, 37, 38, 39, 40, 42, 45, 46, 78, 82, 85, 92, 95, 96, 110, 111, 112, 117, 118
translation start site, 3
triglyceride, xi, 96, 104, 106, 110, 112
tumor, v, ix, x, 17, 27, 37, 51, 52, 53, 54, 55, 57, 58, 59, 60, 61, 62, 63, 64, 68, 69, 70, 71, 72, 73, 74, 75, 76, 78, 80, 82, 87, 99, 106, 116, 124, 132, 136, 137, 142
tumor cell, ix, 52, 53, 63, 68, 69, 70, 72, 73
tumor cell migration and invasion, 68
tumor cell proliferation, 70
tumor necrosis factor, x, 28, 71, 78, 80, 82, 106, 116
tumor necrosis factor-alpha (TNF-α), x, 28, 72, 80, 82, 84, 87, 91, 106
tumor suppressor gene, x, 52
tumor-associated antigen, v, ix, 51, 52, 53, 55, 74, 75
tumor-associated autoantibody, 52, 54
tumor-suppressing gene, 62
tumor-suppressor, 52
turn, 9, 10, 34, 61, 81, 82
type I diabetes, x, 103
type IV collagen 7S, 113
type-1, 133
type-2 diabetes, 133
type-I diabetic state, 108

U

URE-B1, 106
urine, xi, 121, 132, 133, 142

V

vas deferens, 23, 26
vertebrate species, 35
vertebrates, ix, 15, 17, 42
very LDL lipoprotein receptor, 111
very LDL lipoprotein receptor KO mice, 111
vimentin, 69
vitamin C, x, 46, 47, 96, 102, 108, 109, 117
vitamin C deficiency, 96
vitamin C status, 108, 109
voltage-dependent anion-selective channel protein 1, 133

W

Wnt/β-catenin pathway, 70
WT1, 132, 143

X

X chromosome, viii, 1, 3, 18, 19, 41, 104
X-linked gene, 19
Xp12, 3, 4
Xq11.1-12, 3
X-ray diffraction, 18

Z

zinc, 18, 30, 47, 110

Related Nova Publications

USES OF ELECTRICAL STIMULATION FOR DIGESTIVE AND ENDOCRINE SURGEONS

EDITOR: Jaime Ruiz-Tovar, M.D., Ph.D.

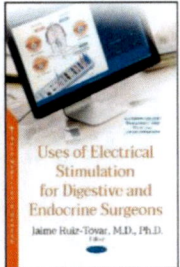

SERIES: Endocrinology Research and Clinical Developments

BOOK DESCRIPTION: The use of electrical stimulators with medical aims has increased exponentially in the last years. The uses are very different. Though the most widely known are referred to as the approaches performed by neurosurgeons, evidence has recently appeared, supporting its use by many other medical specialties.

SOFTCOVER ISBN: 978-1-53615-036-0
RETAIL PRICE: $95

SEROTONIN AND DOPAMINE RECEPTORS: FUNCTIONS, SYNTHESIS AND HEALTH EFFECTS

EDITORS: Monica Munoz and Marshall Mckinney

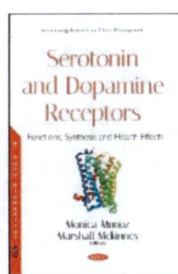

SERIES: Endocrinology Research and Clinical Developments

BOOK DESCRIPTION: In this compilation, the authors begin with a review of the mechanisms of synthesis and secretion, cellular effects and the involvement of serotonin (5-HT) in physiological and behavioral functions horses.

SOFTCOVER ISBN: 978-1-53613-216-8
RETAIL PRICE: $95

To see a complete list of Nova publications, please visit our website at www.novapublishers.com

Related Nova Publications

CYSTIC TUMORS OF THE PANCREAS

EDITOR: Robert Grützmann, M.D., Ph.D.

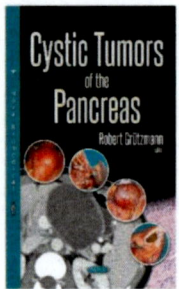

SERIES: Endocrinology Research and Clinical Developments

BOOK DESCRIPTION: Cystic tumors of the pancreas today are diagnosed more frequently in clinical practice, mainly due to an increased use of the modern advanced imaging modalities. Bland cysts of the pancreas most often develop after chronic or acute inflammation of the pancreas. However, the current knowledge concerning the development of cystic neoplasias of the pancreas is still rudimentary.

HARDCOVER ISBN: 978-1-53612-523-8
RETAIL PRICE: $160

PINEAL GLAND: RESEARCH ADVANCES AND CLINICAL CHALLENGES

EDITOR: Angel Catalá, Ph.D.

SERIES: Endocrinology Research and Clinical Developments

BOOK DESCRIPTION: This book presents an overview of the research advances and clinical challenges of the pineal gland. The topics analyzed cover a broad spectrum of functions played by the pineal gland and present new information in this area of research.

HARDCOVER ISBN: 978-1-53612-117-9
RETAIL PRICE: $230

To see a complete list of Nova publications, please visit our website at www.novapublishers.com